I0038018

THE NETWORKED IMAGE IN POST-DIGITAL CULTURE

This collection examines how the networked image establishes new social practices for the user and presents new challenges for cultural practitioners engaged in making, curating, teaching, exhibiting, archiving and preserving born-digital objects.

The mode of vision and imaging, established through photography over the previous two centuries, has and continues to be radically reconfigured by a hybrid of algorithms, computing, programmed capture and display devices, and an array of online platforms. The image under these new conditions is filtered, fluid, fleeting, permeable, mobile and distributed and is changing our ways of seeing. The chapters in this volume are the outcome of research conducted at the Centre for the Study of the Networked Image (CSNI) and its collaboration with The Photographers' Gallery over the last ten years. The book's contributors investigate radical changes in the meanings and values of hybridised media in socio-technical networks and speak to the creeping automation of culture through applications of AI, social media platforms and the financialisation of data.

This interdisciplinary collection draws upon media and cultural studies, art history, art practice, photographic theory, user design, animation, museology and computer science as a way of making sense of the specific cultural consequences of the rapid succession of changes in image technologies to bring the story up to date. It will be of particular interest to scholars and students of visual culture, media studies and photography.

Andrew Dewdney is Co-director and Co-founder of the Centre for the Study of the Networked Image, and Professor of Educational Media at London South Bank University. He has written and lectured widely on new media and museology. His most recent book *Forget Photography* was published in 2021.

Katrina Sluis is Associate Professor and Head of Photography & Media Arts at the School of Art & Design, Australian National University. She is a founding Co-director of the Centre for the Study of the Networked Image and was previously Senior Curator (Digital Programmes) at The Photographers' Gallery, London.

THE NETWORKED IMAGE IN POST-DIGITAL CULTURE

Edited by Andrew Dewdney and Katrina Sluis

Routledge
Taylor & Francis Group

LONDON AND NEW YORK

Cover image: courtesy of Katrina Sluis.

First published 2023
by Routledge
4 Park Square, Milton Park, Abingdon, Oxon OX14 4RN

and by Routledge
605 Third Avenue, New York, NY 10158

Routledge is an imprint of the Taylor & Francis Group, an informa business

© 2023 selection and editorial matter, Andrew Dewdney and Katrina Sluis;
individual chapters, the contributors

The right of Andrew Dewdney and Katrina Sluis to be identified as the authors
of the editorial material, and of the authors for their individual chapters,
has been asserted in accordance with sections 77 and 78 of the Copyright,
Designs and Patents Act 1988.

All rights reserved. No part of this book may be reprinted or reproduced or utilised
in any form or by any electronic, mechanical, or other means, now known or
hereafter invented, including photocopying and recording, or in any information
storage or retrieval system, without permission in writing from the publishers.

Trademark notice: Product or corporate names may be trademarks or registered trademarks,
and are used only for identification and explanation without intent to infringe.

British Library Cataloguing-in-Publication Data
A catalogue record for this book is available from the British Library

Library of Congress Cataloging-in-Publication Data
Names: Dewdney, Andrew, editor. | Sluis, Katrina, editor. | London South Bank
University. Centre for the Study of the Networked Image, sponsoring body.
Title: The networked image in post-digital culture /
edited by Andrew Dewdney and Katrina Sluis.
Description: Abingdon, Oxon ; New York, NY : Routledge, [2022] |
Includes bibliographical references and index.
Identifiers: LCCN 2021061839 (print) | LCCN 2021061840 (ebook) |
ISBN 9780367550585 (hardback) | ISBN 9780367557560 (paperback) |
ISBN 9781003095019 (ebook)
Subjects: LCSH: Multimedia communications–Social aspects. |
Digital images–Social aspects. | Digital media–Philosophy. |
Popular culture. | Mass media and the arts.
Classification: LCC TK5105.15 .N48 2022 (print) |
LCC TK5105.15 (ebook) | DDC 621.382/1–dc23/eng/20220316
LC record available at https://lccn.loc.gov/2021061839
LC ebook record available at https://lccn.loc.gov/2021061840

ISBN: 978-0-367-55058-5 (hbk)
ISBN: 978-0-367-55756-0 (pbk)
ISBN: 978-1-003-09501-9 (ebk)

DOI: 10.4324/9781003095019

Typeset in Bembo
by Newgen Publishing UK

CONTENTS

ILLUSTRATIONS

ACKNOWLEDGEMENTS

This book owes its existence and a debt of gratitude to the research group of the Centre for the Study of the Networked Image (CSNI) in the School of Arts and Creative Industries at London South Bank University (LSBU). In this respect, we would like to acknowledge the financial and administrative support of the School and University over the past decade, which has been essential to keeping our research community alive and growing. In particular we would like to thank Phil Hammond and Elena Marchevska, both staunch supporters of CSNI.

The book is umbilically connected to Martin Lister's earlier project *The Photographic Image in Digital Culture* (1995/2013), in which we were both contributors and whose title we adapt. Martin's scholarly approach to following closely what was happening to the photographic image, focused the attention of a generation of students and researchers and his contribution remains an inspiration to our project.

The book is the product of its contributors, most of whom have been attached to CSNI and whose collective research forms its creative engine. We would like to thank Ioanna Zouli, Gaia Tedone, Nicolas Malevé and Lozana Rossenova for their professionalism and commitment to research as postgraduate researchers. We would also like to thank Geoff Cox and Annet Dekker, current co-directors of CSNI, and Magdalena Tyżlik-Carver, the second CSNI postdoctoral fellow and now visiting researcher for their invaluable contributions. Daniel Rubinstein was a founding co-director of CSNI during his time at LSBU, and we thank him for his thinking, which still reaches the project of this book. Ben Burbridge and Alan Warburton, who we like to think of as fellow travellers of the CSNI, have made important contributions to the book as well as in their collaborations with Katrina Sluis during her time as Digital Curator at The Photographers' Gallery. Past and present CSNI doctoral researchers have generously contributed to research discussions and we would like to acknowledge Nicola Baird, Carolyn Defrin, Jeannete Ginslov,

Victoria Ivanova, Theresa Kneppers, Qian Xiao, Rosie Hermon, Marloes de Valk and Rachel Faulkner.

CSNI has worked with several arts organisations on collaborative PhDs over the last ten years and there are people who have enthusiastically as well as critically supported our approach. In particular we would like to thank Brett Rogers, Claire Grafik, Janice McLaren, Sam Mercer, Jon Uriarte and Jonathan Shaw at The Photographers' Gallery for their long-term support. We would also like to thank Dragan Espenschied and Michael Connor at Rhizome, New York, along with Ben Vickers and Kay Watson at the Serpentine Galleries for their continuing support for collaborative research. At Fotomuseum Winterthur, we would like to thank Marco De Mutiis for his valuable input into this volume in discussion and early drafts, and Doris Gassert for her support and encouragement.

We would also like to acknowledge the support which CSNI researchers have had from the Swiss National Science Foundation and the Post-Photography research group at Lucerne School of Art and Design led by Wolfgang Brückle. The contributions of Katrina Sluis, Nicolas Malevé and Gaia Tedone in this volume develop from their research on the SNSF-funded project 'Curating Photography in the Networked Image Economy' (2018–2020).

Our final thanks are reserved for our families and friends who put up with our obsession.

CONTRIBUTORS

Ben Burbridge is a writer, curator and academic. A former editor of *Photoworks* magazine, his publications include *Revelations: Experiments in Photography* (MACK 2015), *Photography Reframed: New Directions in Contemporary Photographic Culture* (I.B Tauris 2018) and *Photography after Capitalism* (Goldsmiths 2020). He teaches modern and contemporary Art History at the University of Sussex, United Kingdom.

Geoff Cox is Professor/Co-director of the Centre for the Study of the Networked Image at London South Bank University, UK, and Adjunct at Aarhus University. He has a research interest in software studies and contemporary aesthetics, expressed in numerous publications, including Aesthetic Programming (Open Humanities Press 2020) with Winnie Soon, and the forthcoming multi-authored book Live Coding (MIT Press 2022) with Alan Blackwell, Emma Cocker, Alex McLean and Thor Magnusson.

Annet Dekker is an independent curator and researcher. Currently, she is Assistant Professor of Cultural Analysis and MA coordinator of Archival and Information Studies (Media Studies) at the University of Amsterdam. She is also Visiting Professor and co-director of the Centre for the Study of the Networked Image at London South Bank University. She is the author of *Collecting and Conserving Net Art* (Routledge 2018) and *Curating Digital Art. From Presenting and Collecting Digital Art to Networked Co-Curating* (Valiz 2021).

Andrew Dewdney is Co-director and Co-founder of the Centre for the Study of the Networked Image, and Professor of Educational Media at London South Bank University. He was the principal investigator of the AHRC-funded project, *Tate Encounters: Britishness and Visual Cultures* (2007–2010). He has written and lectured widely on new media and museology. He is the author of *Forget Photography* (2021), and co-author of *Post-Critical Museology* (2013) and *The New Media Handbook* (2014/2006).

Nicolas Malevé is a visual artist, computer programmer and data activist. Nicolas has been awarded a PhD on the algorithms of vision at the London South Bank University in collaboration with The Photographers' Gallery. He is a research associate at the Lucerne School of Art and Design and a postdoc at the Centre for the Study of the Networked Image at South Bank University.

Lozana Rossenova is a digital humanities researcher and designer based in Berlin. In 2021, she completed her PhD degree at London South Bank University, in collaboration with Rhizome, working on a redesign of the ArtBase net art archive. Her work focuses on open-source and community-driven approaches to digital infrastructures which organise, store and make knowledge, and different ways of knowing, accessible.

Katrina Sluis is Associate Professor and Head of Photography & Media Arts at the School of Art & Design, Australian National University. She was previously Senior Lecturer and founding Co-director of the Centre for the Study of the Networked Image (CSNI), London South Bank University. From 2011 to 2019 she also held the inaugural post of Senior Curator (Digital Programmes) at The Photographers' Gallery, London, where she is presently Adjunct Research Curator.

Gaia Tedone is a curator and researcher with an expansive interest in the technologies and apparatuses of image formation. In 2019, she completed her PhD at the Centre for the Study of the Networked Image, London South Bank University. Currently, she is working as a research associate at the Lucerne University of Applied Sciences and Arts, and as a lecturer at University Cattolica del Sacro Cuore, Milan, and LABA Art Academy, Brescia.

Magdalena Tyżlik-Carver has a research interest in posthuman curating and computational culture, critical data studies, affective data and data fictions. She is Associate Professor of Digital Communication and Culture in the Dept. of Digital Design and Information Studies at the School of Communication and Culture at Aarhus University (DK). She is a member of Critical Software Thing group and a member of editorial board for Data Browser series.

Alan Warburton is a multidisciplinary artist working with computer graphics to explore ideas of technology, gender, power and representation. His 'hybrid' practice involves operating as a new media artist with international exhibition record, a freelance 3D generalist, and a popular critic of CGI cultures in online video essays. He holds an MA in New Media from LSBU and is a PhD candidate in Film and Screen Media at Birkbeck.

Ioanna Zouli is a researcher and curator with an expansive interest in museums, visual cultures and networked ecologies. Her recent research, in collaboration with the Centre of New Media and Feminist Public Practices, focuses on digital cultural practices and technofeminist methodologies in the Greek context. She has worked with cultural and research organisations in the United Kingdom and Greece, including The Photographers' Gallery, Tate, CSNI, the Royal College of Art and the Onassis Foundation. She was a curatorial fellow of the third SNF Fellowship Programme and Writer-in-Residence at Onassis AiR – School of Infinite Rehearsals: Movement III.

INTRODUCTION

Andrew Dewdney and Katrina Sluis

Worlds

In little over a decade, a combination of advancing technical developments in network computing, Wi-Fi and mobile telephony has cemented a global infrastructure upon which the world's economies and production increasingly depend. Earth is now girded by a network of networked computers, linked to giant server farms, storing incommensurable data, connected by cables, satellites and radio signals, received and relayed between countless devices and the cloud (Bratton 2016). Computing is at the centre of a silent and stealthy industrial revolution (Bridle 2019), extending across an increasingly fragile world ecosystem, touching and reshaping the lives of all living species. How this revolution is understood in terms of the planet, society, government, research and education is now a matter of some urgency. How the Internet and its interfaces are reshaping our sense of self and everyday behaviours is an ongoing matter of vigilance and reflexivity. There is, of course, a need to bring this enormous scale of change into manageable frameworks, to reduce the uneasiness and alarm radical change evokes, and to focus upon positive and practical engagements with networks. Essentially, in education and research, the task for scholarship across the humanities and sciences is to understand, work with, and be critical of the relationships between humans and computers in the spheres in which we interact.

Times

This book is very much about current practices and events related to computational cultures, but its subject, the networked image, has emerged over a period of time, as has academic thinking about the impact of technologies upon media and communication. This gives an historical context to the subject, as well as a

DOI: 10.4324/9781003095019-1

specific window of time over which developments have occurred. Time is itself an important referent in many of the contributions, which attempt either to trace cultural and institutional developments over the recent period or to reflect upon how time is managed in online and software practices; and finally, how time is constituted by algorithms. There is a widespread acknowledgement of the experience of time as accelerating in relationship to the speed of technological change, or of time running out in relationship to the struggle to mitigate the worst scenarios of planetary rises in temperature caused by carbon emissions.

There is also a strong sense, felt in the everyday life of work and leisure, and its patterns of labour and consumption, that the time of the present is cut adrift from the past, which, combined with insecurity about the future, creates a sense of living in a perpetual present (Lipovetsky 2005). It is within these uncertainties about time that current academic research must place itself. Within the humanities, research has been accelerated by an instrumentalised competitive formula of payment by results, based upon reviewed outputs. The time needed to conduct forward looking, open, quality research in the humanities is in the order of a three- to five-year cycle, given the relatively small resources made available to it. In contrast research and development in multinational technology corporations, including those with university links, operates at greater scale and speed with a direct line to application. Research in technology innovation and research in the social, cultural and personal effects of technologies are conducted in separate realms, by different disciplines and institutional arrangements. Historically, this has led to the humanities playing an endless waiting and catch-up game, relegated to commenting after that fact, rather than being conceived as integral to and participating in the actual shaping of a technological future. Commenting upon the social impact of media, art and technology has been the academic path of critical analysis since Marshall McLuhan (1964), Raymond Williams (1974) and others focused upon the post-war expansion of television technologies, through to the emergence of new media in the late 1980s and the work of Donna Haraway (1991), N. Katherine Hayles (1999), Jonathan Crary (1992), William J. Mitchell (1992) and Lev Manovich (2002). It might be said that at the millennial moment, new media theory had reached a level playing field with innovation, offering media practitioners and technical developers' new understandings of the operations of software and their impact upon 'old' media, in a similar way to 1970s semiology offered advertisers even more persuasive ways of selling products. Since then, the pace of innovation and its applications – think here of the launch of Apple's iPhone in 2007 – has accelerated yet again, setting new vistas for media and communications academic research.

This book lies within a very specific timespan of academic publishing, although as we go on to identify, within a different and now transdisciplinary framework from its original photographic frame of reference. Our title deliberately references a previous Routledge book, *The Photographic Image in Digital Culture* (1995, 2013), edited by Martin Lister, which focused on the contemporary image condition, through the prism of photography. Our title signals a passage of time, from the digital to networked image and a further cultural shift from the digital to the post-digital.

We have titled the book on the basis that the term *digital* no longer distinguishes a distinct culture, because the computational is now so enmeshed with what we still call real life. In retrospect we might have retitled the book, 'the networked image in computational culture', or even the 'computational image in post capitalist culture', but this is simply to note the perennial problem of book titles, the important relationship we wish the title carry is that between material technologies and cultural forms.

Lister's two earlier volumes tracked important changes in technical image conditions, across two highly significant decades, straddling the closing of the 20th and the opening of the 21st century. Each decade was marked by a crucial technological and cultural shift. By end of the 1990s, the analogue means of reproduction had all but given way to digitalisation, and by the end of the noughties, the digital had been absorbed by the establishment of computational networked telecommunication. This volume seeks to make sense of some specific cultural consequences of this rapid succession of technological changes and bring the story up to date. The book approaches the task of accounting for the networked image from the perspective of cultural practitioners and the objects and users they are involved with. The perspective of practitioners cannot be emphasised enough in the approach the book takes. It is the contention of the book, written into its fabric, that at this point in time, the nature and complexity of the networked image is best and maybe only graspable through its practices and that while there are useful theoretical accounts, the network remains in many of its practices an uncharted territory. The network image is not easily seen, many of its operations remain veiled, or not legible semiotically, and many more are seen only by machines.

Practices

The book has a further history in originating from research, discussion and debate that has taken place since 2012 at the Centre for the Study of the Networked Image (CSNI), a post-graduate research centre based in the School of Arts and Creative Industries at London South Bank University. For the last nine years, CSNI has been investigating radical changes in the meanings and values of hybridised media in socio-technical networks brought about by new technologies, forms, circulation and patterns of consumption. The centre is unique in its collaborative and embedded approach to researching in real-world contexts where the research questions arise from the developmental needs of partner organisations. CSNI has established partnerships with arts and media organisations which are at the forefront of new media art, including the Digital Programme of The Photographers' Gallery, London, the ArtBase Archive of Rhizome in New York, the Research and Development Platform at Serpentine Galleries, Fotomuseum Winterthur's Digital Programme and the Triangle Network and International Artists Residency programme of Gasworks, London. In addition, CSNI has a network of international visiting scholars whose research is closely aligned to the questions of the book. The distinctiveness of the book is that the chapters emerge from a practical and empirical research culture and from a network of collaborators with closely shared interests. CSNI insists that there is urgency in setting out current developments of visual culture through online curating and machine ways of seeing, the implications of machine learning on imageability, massified image production, machine curation and planetary-scale image-circulation.

Fields

This volume sits at the confluence of several disciplinary and subject streams, the currents of which mingle and merge in the fluidity of the hybridity of networked practices. The key contributory elements, of what might best be thought of as a momentary stilling of the ongoing network flow, congeal around the visual image and the institutions in which images are designed, curated, circulated, exhibited, collected and archived. The book explores the networked image through a transdisciplinary arc of new media practices, media ecologies, art curating, cultural studies, software and computer studies.

Interest in the contemporary condition of the image of course travels, working across the fields of art practice, media practice, art practice, archival practices, curatorial practices as well as studies in art history, cultural and media studies, and the computer vision community. We see the book as located in an emergent field in which the visual and its relationship to technology are subjected to transdisciplinary study. The book draws selectively from these broad contemporary currents, and we would argue their coming together is not by academic design, but by the very forces of the computational mode of production, which academia attempts to describe and keep up with.

Networks

The term *network* has been applied differentially to material and conceptual arrangements connecting and linking components of different orders and types, such that individual elements are defined by a network relation, but this relation is reciprocal in that elements in a network define how a given network is understood. Distinguishing between different conceptualisations of networks is important in this respect. Cybernetics is a primary discourse for thinking about networks. Cybernetics seeks to define human and machine communication in terms of self-regulating and goal-orientated information-processing and therefore has an overarching interest in comparing artificial and biological systems in terms of control systems. Cybernetics privileges the functionality of a network in which any perceived dysfunction would be trained out of the system. Cybernetics defines closed systems, whereas network is used in this volume in open and relative terms, which embrace non-functional, even irrational behaviours. Actor-network theory (ANT), developed in relation to science and technology studies, is a theoretical and methodological perspective to trace connections between what is termed actors/actants and mediators in networks of association (Callon 1984; Latour 1996, 2007; Law 2003). Such an approach informs the working definitions of the networked image taken up in this volume. Latour is at pains to point out that ANT is not a definition of a technical system, such as a rail network, because 'it may have no compulsory paths, no strategically positioned nodes'. Further he removes a second possible misunderstanding by insisting that ANT is not a study of social networks, which are only concerned with human social relations. The definition of networks reached for in this volume is 'framed analytically as an expansive onto-epistemological apparatus, a relational socio-technical assemblage, which both limits and creates possibilities for how and what can be thought/known and imagined within it' (Cox et al. 2021). It is as a socio-technical assemblage that we approach the networked image, as a specific condition in a larger mode of production and reproduction.

Images

Network, as we seek to use the term, as a set of shifting relations between cultural forms and technical systems, is complex enough, but so too is the term image. In the non-representational discourse of digital systems, image has acquired a number of highly specific qualifiers. Academic discussions of the post-photographic have reached for a range of adjectives to define the condition of the image more accurately in networks. The image is variously described as algorithmic (Rubinstein & Sluis 2008, 2013), computational (Beller 2017), operational (Farocki 2004), poor (Steyerl 2009), nonhuman (Zylinska 2019), transactional and even soft (Hoelzl & Marie 2015). Such terms have led to significant insights into what the image is doing on networked screens and contribute to modelling its dimensions, as simultaneously a technological infrastructure and a dynamic of social relations. Undoubtedly images are materially incarnated as well as being present in consciousness as mental images. Images continue to represent, but now the processes of signification are reorganised under a non-representational system of computational capitalism through processes of extraction, abstraction and the financialisation of culture.

The four co-directors of the CSNI research group (Cox, Dekker, Dewdney and Sluis) were invited by Jacob Lund to submit a short paper for the *Nordic Journal of Aesthetics* on 'The Changing Ontology of the Image'. This gave us the collective opportunity to attempt a succinct definition of the networked image, of which the following extract is relevant here:

> The networked image is a cooperation between the quasi-autonomous operations of software and remediated socio-cultural forms. In this sense, the networked image is multimodal and transmedial, and importantly, it presents a specific manifestation of network relations. A networked image emerges through the network; its existence is intricately entangled and intertwined with software, hardware, code, programmers, platforms, and users... We define the image in network culture in three overlapping ways: as a dynamic contingency of human vision; a received historical gathering of material objects embodying that which has been thought and seen; and a new form

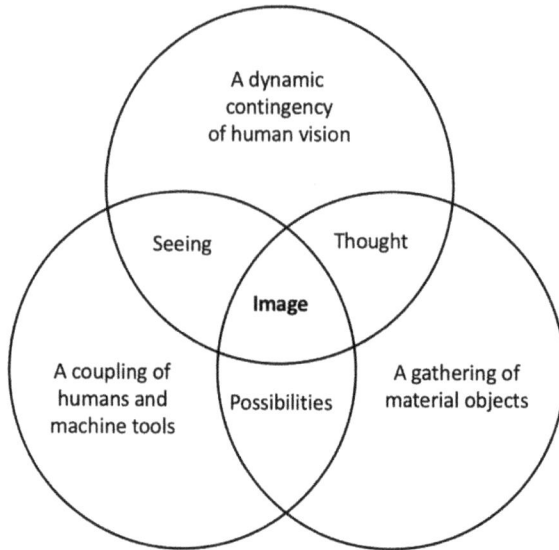

FIGURE I.1 Schematic representation of the networked image

of social relations between humans and machines, in which machines also make images for other machines. In the first sense, the image shapes what is possible to observe and to think; in the second sense, images are material inscriptions of what has been observed; and in the third sense, they are non-representational forms of calculable data.

We concluded the paper by attempting to represent our abstract definition diagrammatically, which we also think has a bearing upon the proceedings of this book:

Knowledges

As already noted, computational capitalism and its expression in post-digital culture carry with it a pronounced sense of temporal dislocation, experienced and managed in highly differential chrono-reflexive ways that consequentially throws life into paradoxical relief. This is nowhere more evident than in the current attempt to manage the deep time of the earth, within a shrinking time horizon. But temporal paradoxes extend everywhere, including the production of knowledge about time and technology in which knowledge contributes to the very changes it attempts to keep up with. This is why we say that the organisation of knowledge and its traditional disciplinary relationships within the university are not outside of his problem and need questioning. The network is in part the result of scientific and technical taxonomies and epistemologies, historically derived from the world prior to the network and which now appear inadequate to cope with the relational nature of knowledge in and of the network itself. Continuing to develop network infrastructures separately from social and cultural knowledge of the effects of networks now appears a high-risk strategy. The conditions of technoscience are for the most veiled by the political and economic terms of research and development and the revolving door between universities and corporate take-up, spin offs and applications. With respect to time and knowledge, it is time to challenge these instrumentalised arrangements.

Post-Digital Culture

The instrumentalisation of knowledge as an outcome of the greater reaches of datafication in all spheres of activity, including the production of knowledge, has been taken up by numerous scholars since Jean Paul Lyotard drew attention to the condition (2004). David Berry and Michael Dieter (2015) take up the problems of what they call the performative logic of the algorithm. They argue that computation has established a constellation of concepts shaping thought and action, or what they more specifically define as an asterism, a discernible pattern within a constellation, of digital instrumentalisation invading cultural and social thinking. For Berry and Dieter, the application of this logic establishes a disjuncture in experience in which separating the digital and non-digital becomes increasingly difficult. Berry and Dieter go on to discuss the relevance of discussion around the terms *post-digital* and *post-internet* as ways of recognising that as the computational infrastructure radiates increased data, it leads to tacit modes of knowing and new iterations of habit. Hence, the idea of defining culture as digital paradoxically excludes the infusion of the digital into the everyday life world and hence opens ways of thinking about the current period as post-digital.

Computational technologies are deeply imbricated in cultural expression, and in this volume, we have lodged the current climate, uneasily as it will become clear, under the rubric of post-digital cultural phenomena, as well as a post-photographic condition, generated by the application of a fully functioning algorithmic and computational network infrastructure. The main aim of this book is to consider recent developments in computational and online cultures to reframe the critical debates on visual culture. Our framing of this volume in terms of 'post-digital culture' seeks to move beyond the binaries of analogue vs digital to understand what happens to the image when computation has become hegemonic. Florian Cramer (2014) explores the utility of the term *post-digital* as a critical reflection on the term *digital*, in which he describes the condition of art and media after the revolution of digital technology as messy and paradoxical. Cramer's examination of the post-digital arises from what he sees as the disenchantment of new media practitioners with the period of fascination with all things digital leading to considerations for critical practice of how to respond and what to do. This is very much the position taken up by the contributors to this volume, who are all seeking answers concerning how to work with the networked image critically and creatively.

Visualities

As we have said, the speed by which computational solutions are developed and applied has created an unprecedented, accelerated, and hybridised media environment in which time is measured as a variable unit of data. It is to this state of hybridity, produced by an assemblage of humans and machines, code and culture, and software and semiosis, which this volume seeks to shed some light upon through the frame of the networked image. Of particular interest across the contributions to this book is the aesthetic of the image in networks, where, in contrast to analogue pictorial aesthetics, the dimension of time is foregrounded through the image's durational, performative and relational characteristics. The network produces culture, the network contains multiple and diverse cultures, and the network is itself a culture. Moreover, the network operates in relationship to the aesthetic practices and objects of older cultural organisations, institutions and practices, the chrono-reflexive terms of which are at odds with the speed of the network. This is true of the university's mode of knowledge production, but it is also distinctly marked in the museological field which operates within historical time. Museums in general have been slow to engage with audiences through the networks, preferring instead to use their online presence as an audience marketing channel. The contributions to this volume, to differing degrees, all address questions of the relationship of the network to curating, exhibition, collection and archives and hence connect with museums and galleries. In what follows, we summarise and draw out the common threads running through the contributions to this volume, threads which we identify as time, scale, labour and relationality. The threads and topics are expressed in Figure I.2.

Organisation of the Book

The book is divided into four parts, organised around four themes related to the overall aims of the book and following something of a narrative logic. *Part I: The*

Topics ⬇	Threads ➡	Time	Scale	Relations	Labour
Part 1 (1-3)	**Reproduction**	Literacy	Capitalism	Social/ Financialised	Individual
Part 2 (4-6)	**Computation**	Algorithm	Data	Operational	Machinic
Part 3 (7-9)	**Curating**	Assemblage	Programme	Hybrid	Collective
Part 4 (10-12)	**Archiving**	Preservation	Digitised	Relational	Care

FIGURE I.2 Topics and threads in the organisation of the book

Condition of the Networked Image aims to reframe the post-photographic debate in terms of the non-representational systems of computing. *Part II: Computation, Software, Learning* sets out how computation is defining new and automated ways of machine seeing. *Part III: Curating the Networked Image* looks at how the networked image circulates in and across online platforms and the emerging cultural forms of value. *Part IV: Digitisation and the Reconfiguration of the Archive* looks at how the networked image is presenting new challenges for cultural institutions concerned with the collection, curation and archiving of cultural value. The book also has four themes or threads running through it, which connect the contributions in different ways. *Time* is a central thread in understanding both the contingency of the networked image, the temporalities of everyday life and the organisation of its knowledge production. *Scale* is also a significant thread in considering the planetary scale of the network itself and within that the technical capacity of producing data at scale. Scale is also an important consideration of the new and extensive forms of human labour involved in producing and interacting with data ontologies and the exponential scale of image circulation. *Labour*, a third thread, is entailed in and by scale and time, but also in terms of the position of labour with waged and unwaged labour of capital forms of production and consumption. In Figure I.1 relations, networks and software really combine in the attempt to describe the networked image as a complex assemblage and contribute towards a meta-view of defining features of a new paradigm of the visual in culture.

Part I: The Condition of the Networked Image

The labour involved in the production/reproduction of the networked image forms one of the book's significant interests. In *The Politics of the Networked Image: Representation and Reproduction* (Chapter 1), Andrew Dewdney recognises the need to develop a politically motivated analysis of the default image produced by computation and its relation to the world, based upon understanding both technical network operations and the organisation of capital and labour. He approaches the networked image in three related ways in order to rethink visuality and what happening to ways of seeing. Like most of the authors represented in this volume, Dewdney defines the image as a relational assemblage, combining human and machine behaviours, in which the image can be thought of as a transaction in a data network. However, Dewdney considers this an insufficient account of the forces operating within and reproduced by network infrastructures. Building upon the idea of the network as a financialised data system, Dewdney considers the image as a form of the social relations of the reproduction of labour. But even this socio-economic definition, overlaid upon the technical assemblage of the networked image, is not considered sufficient in gaining an overview of the paradigm shift in visuality he describes. In a third move, Dewdney considers the image in relationship to the socio-historic symbolic realm of the circulation of what he terms *transmedial* heritage. The transmedial operates in the temporal space of networks, software and the programmable image forming narrative flows and transactions in the expression and regulation of subjectivities. Dewdney concludes with a question and a call, asking: how can a new discipline of knowledge of the visual be constructed to traverse the dimensions, the peaks, and troughs of the global, algorithmic, telecommunication landscape, which includes its infrastructure, operations, interfaces, tools, hybrid images and user behaviours? This leads Dewdney to the view that, 'Such a project in the future will be able to articulate the reality the networked image produces and is productive of. The call for new forms of computational literacy is in the final analysis a form of political literacy'.

In *The Networked Image after Web 2.0: Flickr and the 'Real-World' Photography of the Dataset* (Chapter 2), Katrina Sluis examines the circularity of image circulation as a means of tracing the shifting socio-technical context of the image after Web 2.0. She does this through an account of the rise and fall of Flickr as a lens through which the ubiquity of photography might be glimpsed, and more recent developments of data capitalism observed. Sluis details how the historic snapshot, which formed the basis of online image-sharing communities, became the basis on which modern machine learning systems recognise and categorise the world. However, for Sluis, this is an elliptical process because to determine the value and relevance of photographs which circulate beyond the limits of human attention the image demands algorithmic classification. Yet at the same time, the classificatory powers of machine learning depend upon the creative labour of the users of photo-sharing platforms who produce images.

Sluis together with Daniel Rubinstein (2008, 2013) developed the idea of the networked image as a new paradigm of the authorless, contextless, screen-based image. In their work Sluis and Rubinstein connected the networked image's agency to the massification of the snapshot, which was subject to the centralising conditions of the database-driven web, where visibility depends on human and nonhuman agents. In this volume she argues the networked image is the manifestation of a non-representational mode of image production, in which, 'the iconicity of the image [is] being hollowed out and declining as a site of cultural value, whilst the relation between images became more valuable'. From this perspective the photograph on-screen is an output of algorithmic processing, in which the photographic image is positioned as a cultural residue of human–computer relations that lay beyond the screen. In 2013, as a result of their investigations into the algorithmic image, Sluis and Rubinstein called for a rethinking of the image as processual and an outcome of software, which they argued would require a new vocabulary to address the new image economy. It is precisely this call, which led to the formation of the CSNI and, a decade later, this volume which details some of the outcomes of this research. In her chapter, Sluis returns the argument of the algorithmic image to the events and developments of the last decade as a means of understanding how the aspirations of the creative amateur, as producer of user-generated content, have been adapted. Now the hopes of the image-sharing community have been converted into the extraction of data in platform capitalism and the financially empowered influencer. Sluis offers us an important advance on the conclusion reached in her 2013 paper with Rubinstein, where she defined the main characteristic of the algorithmic image is its undecidability. In her contribution to his volume, Sluis identifies how a photographic logic is coded at both the backend and frontend of the screen image through the use of image datasets in which the snapshot is considered a ground truth of a 'real-world' scene by computer vision. Sluis says, 'the networked image vacillates between fluidity and categorisation, between liquidity and classification, between the amorphousness of the big data cloud and the specificity of someone's actual cat'. This leads her to the insight that for the networked image to achieve semiotic clarity, it requires curation as a necessary condition of its stabilisation. In reaching this position, she passes the baton to Nicolas Malevé whose research on the social ontology of computer vision opens up new ways of understanding how, in Sluis' terms, 'the photographic pipeline of machine learning seeks to stabilise and align the inherent polysemy of the image in order to produce actionable insights'.

In *Post-Capitalist Photography* (Chapter 3), Ben Burbridge goes some way to answering Dewdney's call for a political literacy, although he does so by embracing the persistence of photography in post-capitalist terms. Burbridge's goal here is to think about the political possibilities for the commoning of visual culture, which socio-technical shifts in image culture make possible. Burbridge defines post-capitalist photography as a method, an idea and a speculative future, which he names as a form of radical democratic collective subjectivity. Burbridge recognises that post-capitalist photography articulates a 'profound asymmetry', between the principles of sharing which characterise the social production of images, against the extractive, monetised interactions of data harvesting. Burbridge's argument leads him to suggest the possibility that collective experience within the photographic universe will be more powerful than ideology and reveal the plutocracy driving the financialisation of image labour. For Burbridge, the social collectively inherent in image sharing is a possibility opened up by post-capitalist theory, which argues that with increased automation, capitalism will have to adapt to an economy which pays a universal social wage in order to maintain its cycle of production and consumption. Burbridge argues that such egalitarian and collectivist possibilities are emerging at the moment when photography is reconfigured by networks; however, he also sees that at the same time, the material realisation of egalitarian potential is facing unprecedented challenges. The novelty of Burbridge's view of contemporary image culture is to see that it is founded upon the contradiction of unpaid labour, a situation which might alternatively call for wages for Facebook, unionisation, opting out of social media or developing alternative platforms. Yet Burbridge finds such traditional forms of action unlikely. Instead he sees the social production of value within the monetisation of human interactions, not encompassed by previous industrial forms of labour. The relationship between labour and photography can be understood in terms of the continuous demands on capitalism, transformations in which informational labour, which carry a greater degree of autonomy and authenticity in a 'flexitime world where success is gauged not through movement up definable hierarchies but by a capacity to move between projects and across global networks'. Burbridge's insistence on photography as labour in the reconfiguration of industrial labour directs attention beyond the image to illuminate other forms of work, upon specific tasks and the conditions under which they are performed, thereby heralding the possibilities of social collectivity inherent in networks. In conclusion Burbridge argues that post-capitalist photography performs an important task. It can map the inequities of a neoliberal world through the networked image, situating cognitive and manual labour in direct relation to each other and confront the asymmetry that currently defines photography's interactions with the network.

Part II: Computation, Software, Learning

Time and scale are centrally involved in Nicolas Malevé's *The Computer Vision Lab: The Epistemic Configuration of Machine Vision* (Chapter 4). Malevé undertook a four-year research project which involved the re-staging of a 2007 experiment conducted by computer scientists at California Institute of Technology and Princeton, which studied what can be perceived from a digital photograph in a millisecond glance (Fei Fei et al. 2007). The experiment informed the development of ImageNet, a canonical dataset of 14 million 'photographic images' which have been scraped from the web, cleaned and labelled before being used as training data in the field of computer vision (Deng et al. 2009). Malevé quotes ImageNet's creator, Professor Fei-Fei Li, who jokes in a talk that because of the enormous scale of ImageNet's processing, it would take one of her graduate students, forsaking sleep and looking at approximately two images per second, over 19 years to view them and detect erroneous results. Instead, the verification and labelling of the dataset was outsourced to Turkers, workers on the Amazon Mechanical Turk (AMT) platform, performing what Amazon terms *Human Intelligence Tasks* (HITs) – micro-tasks where workers are paid at low rates, for example $0.01 per HIT, and must therefore label images in small fractions of time to maximise their wages. As Malevé points out, 'in cognitive psychology, time is a crucial factor in visual perception in which visual perception is understood as an act that unfolds in a particular temporal context'. Fei-Fei Li's method of producing the training data for machine vision algorithms is adapted from techniques used in cognitive psychology for studying how the eye and brain process visual signals, in which 25 milliseconds is considered a threshold for consciousness to operate. Malevé's careful re-experiments of vision at speed, carried out with students and the public at The Photographers' Gallery, London, allowed him to analyse what he describes as the 'enfolding of the computer vision lab and the annotation platform' and their increasing imbrication 'with each new cycle of investment and innovation'. His concern was to approach machine vision as more than mathematics and code, the popular view of what computer scientists do, and instead to understand machine vision as distributed, related to a larger field of programmability spawning endless cycles of image capture, annotation, aggregation in datasets, experimentation, classification, measurement, investment. The political implications of Malevé's work are that the model of machine vision currently in operation privileges an understanding of the world at a level of abstraction that requires the least cognitive effort. Further, this model did not stay in the lab but required an industrial level of labour, which is managerial as well as epistemic. As

Malevé concludes, to change the direction of computer vision would mean more than coding a different algorithm. It would require an 'intervention in the social ontologies of machine vision, in its division of labour and the relations between its research institutions and its industrial environment of production'.

The networked image, understood as a dynamic, distributed and computational object, unsettles received notions of space-time. What emerges is that time is constituted through the relational and processual nature of the networked image, in which the dynamic of social relations as much as technological infrastructure is involved. One dominant form of those social relations is time as labour power. Burbridge's aspiration for a post-capitalist photography with which to map the inequalities of the neo-liberal world is reframed by Geoff Cox's account of the need for a post-digital literacy, which he sees as the complex task of understanding machine ways of seeing. As Cox suggests in *Ways of Machine Seeing as a Problem of Invisual Literacy* (Chapter 5), 'If we want to *see* the invisible world of machinic visual culture, we need to unlearn how to see like humans and learn to see more like machines'. How he arrives at this conclusion is through a discussion of the kinds of literacy such a task would entail. Importantly Cox compares the project of machine ways of seeing to John Berger's enduring 1977 television series and book, *Ways of Seeing*, which serves as the backdrop to his own reflexive approach to computation. As he puts it, 'When images are made by machines for other machines, what kind of literacy applies, if at all?' The comparison with Berger's *Ways of Seeing* enables Cox to articulate the ambiguities between seeing and language and insist upon the limits of structuralism and semiotics to account for the networked image. If images and signs are now part of the operations of capitalism itself, distributed across networks, in which images are now made by machine for other machines, images no longer represent things in the world, but as Cox puts it, 'are an active part of invisible visual culture'. What kind of visual literacy, of computational literacy, is needed to understand that it is no longer a question of simply looking at an image, but rather that the image is looking at us? Again, as Cox puts it 'in the case of computer vision systems, they make judgements and decisions, and as such exercise power to shape the world in their own images, which, in turn, upholds the argument that they embody new ways of seeing'. Cox goes on to assess computer literacy in terms of coding and programming language, recognising that the proliferation of procedural literacy is predominantly a managerialist pedagogy, encouraging us to think in the logistics of capitalism. However, rather than dismissing programming literacy as serving only needs of policy-makers, Cox suggests that a literacy is needed now more than ever to understand ways of machine seeing as a politics of experience and value. The development of a post-digital literacy would involve, 'developing a literacy that is co-constituted, one that is more sensitised to relational operations and that shifts our attention away from the acquisition of technical know-how alone to new possibilities for aesthetic practice'. Such an approach is not a matter of the acquisition of technical know-how, but one in which know-how is opened up to social imaginaries. He concludes, 'If we want to *see* the invisible world of machinic visual culture, we need to unlearn how to see like humans and learn to see more like machines, or rather, see like both in ways that departs from a Western-centred humanist standpoint and thereby embrace intersectional methodologies'.

Thinking like a machine, or rather, understanding and working with the language of new media, or in the cases considered here, using software is a further way of looking at the parameters of the networked image. For the contributors to this volume, politics is written into software. The recognition of the value-loaded, political, nature of computational systems leads the contributors to emphasise the importance of *theory-in-practice*, as a necessary component of critical analysis and as a means of resisting and developing alternative practices. What this entails for a new media practitioner is taken up in Alan Warburton's contribution *Soft Subjects: Hybrid Labour in Media Software* (Chapter 6), in which he re-examines Lev Manovich's (2001) concepts of 'soft evolution' and 'permanent extendibility' in relationship to new media practice and concludes by arguing for the usefulness of thinking of the practitioner as the 'soft subject'. Warburton starts by recognising that today commercial software product ecosystems are tightly and inextricably locked into most professional media practices, regardless of scale or industry. He points out that while the critical lineage of early 2000s net art continues, commercial software now prevails and most open-source alternatives deviate little from the schematics of their commercial counterparts. This leads Warburton to argue that:

> This forced conversion of relatively inflexible analog wholes to *flexible*, *editable* and *configurable* digital parts seems to be the evolutionary survival principle of metamedia, the superpower that determines why new media 'computational' ontologies overtake old media 'cultural' categories, why they continually produce change within software ecosystems and supply a sequence of new standardised file formats, applications, plugins, devices and workflows.

For Warburton, the critical reflexivity he seeks in order to confront such complicity will be found by attending to the forms of subjectivity, labour, thought and action of the 'subjects who embody a complex contemporary sociotechnical milieu'. Warburton characterises the last two decade as marked by recursive cycles of digital disruption and integration in which practitioners know well the political realities of working in digital practice, but that any politics becomes subordinate in the challenging software environment of 'marshalling constantly changing technical systems to produce spectacular or novel creative artefacts'. Warburton, quoting Matthew Fuller (2006, p. 9) recognises software labour as the exploitation of capacity of the 'general intellect' to 'take bits from here and from there, to recompose multiply encoded and gated, broken, esoteric and public materials and information and make something of them'. For Warburton the conditions which keep criticality at bay are lamentable, given that those involved in visual media are 'at the frontline of many of the most controversial technical advances in image production, which usually involve machine learning, image recognition and image synthesis'. Although it is the deregulated working conditions which are primarily responsible for the divorce of theory and practice, Warburton also points to important limitations of theory to push beyond a technical account of media, recognising that Manovich was right to define the mediality of software as permanently extendable, remixable, hybrid and modular; it does not include the subject as an active component within media. This is where Warburton introduces his notion of the soft subject as a counter to the limits of technical soft evolution. Media professionals today perform various kinds of immaterial, cognitive and network labour, but critical attention to affective labour, indicative of 'playbor', gamification and crowd work, is hampered by the illegibility of complex media software which the 'soft subject' encounters in practice. Warburton concludes by arguing that the 'connective tissues' of metamedia 'means treating every media assemblage as a distinct instance of contingent technical evolution', in the vein of Bruno Latour's concept of hybridity. What has been underestimated in Warburton's view is that the labour of the soft subject, 'must recursively metabolise obsolescence by pioneering new forms of computational control, overcoming path dependence and habituating to high levels of technical debt'.

Part III: Curating the Networked Image

In *The Paradoxes of Curating the Networked Image: Aesthetic Currents, Flows and Flaws* (Chapter 7), Gaia Tedone takes her online curatorial practice as a starting point for considering the dynamics of online image searching and the circulation of images within computational culture. Tedone also notes the significance of time in the rapidity with which images circulate across the Internet, with the consequence of multiplying their contexts of reception and patterns of interpretation. The rapid 'flow' of images leads her to explore the concept of circulation, which she sees as 'the link between the operations of the networked image and the function of online curation' and a way of 'framing the politics and aesthetics of curating the networked image'. For Tedone, online curating engages with the massively distributed terms of the image in which users engage in curating through the organisation, filtering and arrangement of digital images and information. However, Tedone goes further than noting the interactions of users in the distribution of images, recognising that circulation entails algorithms and software embedded in social media platforms, which are used in the processual relay of the image, in which they operate as data. As such, Tedone concludes that the operations of the networked image are excessively performed through the actions executed by digital professionals, online users and software, which applies to 'any sellable product or service [that] can now be virtually curated'. Perhaps more significantly, these forms of content curating, supported by a range of digital data aggregation tools and software, are in Tedone's view, shaping the professional field of 'the 21st century knowledge worker'.

Ioanna Zouli's chapter, *Internet Liveness and the Art Museum* (Chapter 8), discusses Tate's excursion into distributing performance art live and online and articulates the hesitancy and suspicion of art museum curators towards the unknowns of network culture and Tate's subsequent retreat to established art museum conventions and control of content through an edited broadcast form. Her case study of Tate Modern's early *BMW Tate Live: Performance Room*, a strand of 'live' programming which took place over 2012–2015, shows the art museum's difficulty and puzzlement in attempting to manage the instantaneity of audience engagement in a live YouTube performance, which highlights a series of misunderstandings in the transactions between curator, artist and audience operating in the online environment of the YouTube platform. Zouli's careful analysis of Tate Modern's *Performance Room* project illuminates several difficulties in embracing 'realtime' and hence 'liveness' on the YouTube platform. The liveness of the art performance in the real time and space of the gallery places what the audience is watching on screen and their 'live' responses in a very different relationship to the performers. Further, as Zouli explains, the subsequent invited discussion on the performance could not be moderated within the conventions of the physical presence of the audience, curators and performers. Zouli's examination of the tensions involved in art stepping into a commercial online platform is highly pertinent to the situation of the last two years of the global Covid-19 pandemic, in which many cultural organisations migrated parts of their programme online. What is notable in this is the extent to which organisations are still largely adopting an analogue model of broadcasting curated or edited content, rather than exploring the possibilities of dialogic interaction.

In *Screenshot Situations: Imaginary Realities of Networked Images* (Chapter 9), Magdalena Tyżlik-Carver comes to see the operations of the networked image through an examination of the screenshot, which she argues has been overlooked, given the contemporary ubiquity of screens, with which the image has become synonymous. She makes the point that because the infrastructures of screenshots are not visible as part of the computational infrastructure, it is necessary to analyse screenshots as connected to the socio-technical conditions of their making. Tyżlik-Carver traces connections between the scientific and vernacular through the heterogeneity of 'screenshots as always scientific and technical, while also social and aesthetic'. And it is through these very logics, Tyżlik-Carver argues, that the networked image is revealed. This is the screenshot not experienced as a still image on screen, but as a unit of measure which accounts for the apparatus that cannot be seen in the execution and performance of code based on human and machine collaborations. Tyżlik-Carver observes that 'screenshots are situations that capture the circulation of data at the moment when data materialises as an image' in which 'real conditions of screen capture become an image'. Unpacking the screenshot, Tyżlik-Carver argues, makes it possible to specify what the networked image is, which for her is not a new category of image, but rather a located agential force, 'a situation that organises relations and bodies that come together to make an image', including a human operator, the machine, as well as tools facilitating input and output.

Part IV: Digitisation and the Reconfiguration of the Archive

In *Networks of Care* (Chapter 10), Annet Dekker shifts attention to archival and preservation practices, specifically of born-digital objects. The chapter asks how the network impacts upon conventional preservation methods, raising the additional question of how the future's past will be filtered through the networked image. In the same manner as Tedone, Dekker sees networks of care as processual, as a continuously reconfigured object of preservation, connected in memory, space and time. Dekker emphasises the dynamic of caring in terms of an 'assemblage of contemporary practices, platforms, software, computer programs and human hardware'. Here again the network and its meanings are seen through its production and circulation, which in networks of care happens over time as well as simultaneously the 'real-time' of networked data processing. The processual nature of care means that the born-digital object is always in a state of emergence through the relations of software, hardware, code, programmers, platforms and users. The defining aesthetic of networks of care shares a close resemblance to the networked image as both emerge from the relational assemblage of art, technology, politics and social relations. Dekker is primarily concerned with the practices of digital preservation and, through her analysis of the relationality of digital objects, directs attention to preservation models which consider the socio-technical relations and negotiations that are necessary to produce and maintain digital-born objects. In digital preservation practices, Dekker translates the dimension of time in networked operations to the dimension of care within networks in which attention is directed at the open, unexpected and unstable characteristics of assemblages. Finally, Dekker concludes that preservation of born-digital objects 'can be understood as a speculative practice, where knowledge unfolds between subjects (human and non-human) whose ability to know is mediated by how they reach out, and by the receptivity of the other'.

What is practically and technically involved in the preservation of digital-born objects as relational assemblages, as Dekker defines them, is examined in closer detail in Lozana Rossenova's, *Beyond the Screenshot: Interface Design and Data Protocols in the Net Art Archive* (Chapter 11). Rossenova's analysis of the conceptual and practical dimensions of the preservation of born-digital objects has a wider importance for the book in understanding the operations and hierarchies of data and software interfaces entailed in the networked image. Rossenova's introduction starts by noting that over the past two decades, the proliferation of scanning technologies, digital record keeping and access to cheap manual labour have led to the exponential growth of mass-digitisation and online archives and collections. Few, if any, international museums can be without a digitised version of its collection and Google Arts & Culture has built its own considerable digitised archive of international artworks under the mission statement of making cultural heritage accessible to all. As Rossenova observes, for most, this access is to scan text pages or photographic images of analogue objects operating within a representational framework of the photographic paradigm, as fixed and static objects. Technically, institutions managed digital objects in data asset repositories which 'ensure that the files remain unchanged at bit level and attendant metadata schemas ensure images are preserved as fixed entities, and *not* as variable, networked data'. As Rossenova goes on to say,

> Institutional protocols usually prevent design research projects from deeper engagement with infrastructure – projects tend to remain 'client-side' or frontend-focused, i.e. pertaining to what users see in their own browser, which is the 'client' software requesting data from the server software, maintained by the institution.

In exploring the preservation of net art, Rossenova observes that the current state of digital preservation practice becomes increasingly untenable, in the face of the need and desire of cultural institutions to collect complex, non-linear and networked born-digital cultural expressions.

Rossenova, a trained UX (User Experience) designer, produced a redesign of the ArtBase archive over the course of a three-year research project. Her research examined the modelling of relations between the operations of the database, which structures archival data (the backend) and the interface, which enables user interaction (the frontend). Rhizome is a New York-based digital arts organisation founded in 1996 which is dedicated to promoting born-digital art. The ArtBase archive project, founded by Mark Tribe in 1999, now comprises over 2000 net art artworks spanning a 20-year history. Rossenova argues that net art, and other forms of born-digital culture, can be understood as a particular form of networked, or even co-networked, relational images, which challenge established approaches to cultural heritage archiving and preservation. The problem at the outset, for redesigning the ArtBase archive, lays in the fact, as Rossenova explains, 'established metadata standards, collections management software and end-user interfaces cannot account for the needs of temporally variable and performative digital artifacts with networked technical dependencies and user interaction requirements'. This is a problem already recognised by curators, conservators and researchers working with net art collections of the need for preservation tools which can capture the processual, performative and variable properties of born-digital artworks. The insights afforded by Rossenova's study is that archival records can no longer understood to be a static, value-neutral entity, but rather a dynamic process of production and interpretation, carried out by multiple agents. This leads Rossenova to argue that there is a 'need to rethink the image-as-artwork-representation paradigm and expand the notion of the image beyond static representation towards operative, data-rich environments'. Precisely because of the interdependency of interface and user, Rossenova's concludes that 'the problems of the networked archive, are not just archival science problems, or design problems, or computer science problems, but problems of relations across diverse entities and agents'.

Critical Perspectives

The contributions to the volume are located within a broadly materialist critical tradition of cultural, art, media and communications studies, which understands image formation as the outcome of complex socio-economic technical practices. The emphasis upon the materiality and technics of the network apparatus has opened up new ways of thinking about the agency of machines and their interactions with humans, but it also leads back to the need to rethink the position of visualisation in the general modes of cultural reproduction and representation. It is in this sense that contributors draw upon contemporary Marxism and non-foundational process and constructivist theories from science and technology studies, ethnographies and feminist technoscience. Whilst the arguments emerge from difference practices and discourses, taken collectively, they articulate a shift in ways of understanding the networked image as a complex, socio-technical assemblage in which humans and machine agency is entailed and through which contemporary forms of power operate. Western scientific conceptions of universal thought posit a singular ontological and epistemological vantage point defining the world has been challenged from a number of quarters in post-human (Braidotti 2013) and feminist technoscience theory (Haraway 1991, 2016). The influence of Karen Barad's important book, *Meeting the Universe Halfway: Quantum Physics and the Entanglement of Matter and Meaning* (2007), is notable in more than one chapter in theoretically framing the human-nonhuman relationship in computation in terms of a relational intra-action. Such concepts have a bearing upon the contributions to this volume in terms of regarding the networked image as a socio-technical assemblage, which has global reach.

References

Beller, Jonathan. 2017. *The Message is Murder: Substrates of Computational Capital*. London. Pluto.

Berry, David M. and Michael Dieter. 2015. "Thinking Post Digital Aesthetics: Art, Computation and Design." in *Post Digital Aesthetics: Art, Computation and Design*, edited by David M. Berry and Michael Dieter. London. Palgrave/Macmillan, pp 1–11.

Braidotti, Rosi (2013) *The Posthuman*. Cambridge. Polity.

Bratton, Benjamin (2016) *The Stack: On Software and Sovereignty*. Cambridge, MA. MIT Press.

Bridle, James (2019) *New Dark Age: Technology and the End of the Future*. London. Verso.

Callon, Michel (1984) "Some Elements of a Sociology of Translation: Domestication of the Scallops and the Fishermen of St Brieuc Bay." *The Sociological Review* 32 (1) (May 1984): 196–233. https://doi.org/10.1111/j.1467-954X.1984.tb00113.x

Cox, Geoff, Annet Dekker, Andrew Dewdney and Katrina Sluis (2021) "Affordances of the Networked Image." *The Nordic Journal of Aesthetics* 30 (61–62): 40–45.

Cramer, Florian (2014) 'What Is "Post Digital"?'. *APRJA* 3 (1). https://aprja.net//issue/view/8400/893

Deng, Jia, Wei Dong, Richard Socher, Li-Jia Li, Kai Li, and Li Fei-Fei (2009). "ImageNet: A Large-Scale Hierarchical Image Database." In *2009 IEEE Conference on Computer Vision and Pattern Recognition*, 248–255. https://doi.org/10.1109/CVPR.2009.5206848.

Farocki, Harun (2004) "Phantom Images." *Public* 29: 12–22. https://public.journals.yorku.ca/index.php/public/article/view/30354/27882.

Fei-Fei, Li, Asha Iyer, Christof Koch, and Pietro Perona (2007) 'What Do We Perceive in a Glance of a Real-World Scene?'. *Journal of Vision* 7 (1): 10. https://doi.org/10.1167/7.1.10.

Haraway, Donna (1991) *Simians, Cyborgs and Women: The Reinvention of Nature*. London. Free Association Books.

Haraway, Donna (2016) *Staying with the Trouble (Experimental Futures): Making Kin in the Chthulucene*. Durham. Duke University Press.

Hayles, N. Katherine (1999) *How We Became Posthuman: Virtual Bodies in Cybernetics, Literature, and Informatics*. Chicago, IL. University of Chicago Press.

Hoelzl, Ingrid and Rémi Marie (2015) *Softimage: Towards a New Theory of the Digital Image*. Bristol. Intellect.

Latour, Bruno (1996) "On Actor-Network Theory: A Few Clarifications." *Soziale Welt*. 47 (4): 369–381. Nomos Verlaggesellschaft mbH.

Latour, Bruno (2007) *Reassembling the Social*. Oxford. University Press.

Law, John (1992) "Notes on the Theory of the Actor Network: Ordering, Strategy and Heterogeneity." *Systems Practice* 5 (4): 379–93.

Lipovetsky, Gilles (2005) *Hypermodern Times*. London. Polity.

Lister, Martin (ed.) (2003) *The Photographic Image in Digital Culture*. London. Routledge.

Lyotard, Jean-Francois (2004) *The Postmodern Condition: A Report on Knowledge*. Manchester. Manchester University Press.

McLuhan, Marshall (1964) *Understanding Media: The Extensions of Man*. New York. McGraw-Hill.

Manovich Lev. (2002) *The Language of New Media*. Cambridge, MA: MIT Press.

Matthew Fuller. 2006. *Softness: Interrogability; General Intellect; Art Methodologies in Software*. Aarhus: Center for Digital Æstetik-forskning.

Mitchell, William J. (1992) *The Reconfigured Eye: Visual Truth in the Post-Photographic Era*. Cambridge, MA. MIT Press.

Rubinstein, Daniel and Katrina Sluis (2008) "A Life More Photographic: Mapping the Networked Image." *Photographies* 1 (1): 9–28. https://doi.org/10.1080/17540760701785842.

Rubinstein, Daniel and Katrina Sluis (2013) "The Digital Image in Photographic Culture: Algorithmic Photography and the Crisis of Representation" in *The Photographic Image in Digital Culture*, edited by Martin Lister. London. Routledge, 22–40.

Steyerl, Hito (2009) 'In Defense of the Poor Image'. *e-flux Journal* 10 (9): 9 https://www.e-flux.com/journal/10/61362/in-defense-of-the-poor-image/.

Williams, Raymond (1974) *Television, Technology and Cultural Form*. London. Fontana.

Zylinska, Joanna (2019) *Nonhuman Photography*. Cambridge, MA. MIT Press.

PART I

The Condition of the Networked Image

1

THE POLITICS OF THE NETWORKED IMAGE

Representation and Reproduction

Andrew Dewdney

More than Visual

This is the time of the networked image, which is transforming settled ways of seeing, cemented over the course of the 20th century by the analogue means of visual reproduction in photography, film and television. The socio-historic code of the analogue technical mode of reproduction remains that of representation, a system of symbolic equivalences, based upon inscription, resemblance and recognition, a mode which ensures a naturalised, if not ideological reality, epitomised visually by the photographic image. With computing, the surface of the screen simulates the image of representation, as a necessary interface to the operations of data and signalling. In the first period of digitisation, across the late 1980s and 1990s, the digital image was initially understood in the analogue mode, as a static electronic version of the photographic image. But over the last two decades, through increased capacity of computing and scale of data, the new default of the image is its position in the processual relays of the network of networked computers. This change from digital to networked image has occurred because, while the front end of computing, what Lev Manovich (2001, p. 45) termed *the legible cultural layer* is still realised as representation, the backend of computing, the illegible computer layer, has developed an infrastructure on an industrial scale, serving an information economy. The near-limitless production of data in networks drives towards ever greater automation in which computers communicate without the intervention of humans. From the perspective offered here, the operations of computational networks, the black boxing of neural networks, the configuration of algorithms in software applications, proprietary platforms and mobile devices, call for a radically new understanding of what images are doing to us and what we are doing to and with images.

DOI: 10.4324/9781003095019-3

For the academic disciplines in which the visual image is studied and practiced, the computational networked image presents both new problems and new opportunities in revising theories and practices of visuality. The widespread use of computational 'tools' demands a re-evaluation of established institutional, historical and ontological categories of the visual in fine art, media, architecture, design, graphics, illustration, cartography, film, photography and animation. The perspective offered here suggests that visuality has entered a decisive era of the *more than visual* and *nonrepresentational* in which an ocular-centric worldview, which previously devolved upon the mechanical eye, has been overturned by the operations of data and signalling. The ways of seeing, established by the European Enlightenment and its humanist tradition, have been rewired by computation and now demand new ways of thinking about visuality, temporality and the human sensorium. This is a situation in which Hito Steyerl (2013) can describe images as crossing the screen and acting in reality. It is a situation of simulation argued by Jean Baudrillard (2012, p. 20) in which images no longer operate as mirrors to reality, or as reflections of inner mental states, but are projected directly, without mediation into reality. It is a situation in which the semiotics of the sign, the signifier and signified no longer easily apply. Or, as Deleuze and Guattari (2004, p. 163) think about the same problem, a situation in which language no longer signifies something which must be believed. The assumption of a transparent representational relationship between the image and reality is assailed on all sides, by the technical apparatus of nonrepresentational mathematical code, by intellectual doubt about the status of universal scientific truths, and by right-wing populism and fake news (Latour 2018). As a consequence of such perspectives, the argument pursued here is that it is no longer politically progressive, nor useful, to study media images, in whatever form, as separated from the world historical events, materiality, epistemologies, politics and experiences they participate in.

As this chapter recognises, developing a politically motivated analysis of the new default image of computation and its relation to the world necessarily involves more than one strategy, but where a synthesis is not easily achieved. What follows frames the networked image in three related ways, in the hope that they are consistent with the current problems of rethinking visuality, as well as contributing to other attempts to map out a thorough-going overview of what is happening to ways of seeing. Firstly, it considers the image as a relational assemblage, combining human and machine behaviours, in which the image is thought of as a transaction in a data network. Secondly, building upon the idea of the network as a financialised data system, it considers the image as a form of the social relations of the reproduction of labour. Finally, it considers the image in relationship to heritage, temporality and how the programmable image and the globalised network lead into the state of trans-hypermedia.

There has been a noticeable quickening of academic knowledge concerning the status of the image over the last decade, such that understanding is finally keeping pace with the magnitude of image capture, its aggregation in datasets, cloud storage, platform interfaces, circulation and deployment in machine vision.

This chapter sets out, albeit briefly, to sketch what it considers to be some of the most cogently argued understandings of the operations of the network image. But its deeper purpose is to consider the political implications of this knowledge in everyday life, including the conditions of its own production, as well as its opening out on to a revised version of the material state of the world. If understandings of media are changing dramatically, so too are understandings of the world. It is increasingly recognised, certainly to readers of this volume, that individually and collectively, we, the global human population, are living through a profound disturbance of the planetary ecosystem. This situation confronts everyone in the attempt to understand the human impact on the planet and, closer to the interests explored here, confronts media and cultural scholars with how the humanly constructed world is sensorially mediated. We might also add that all living species are in one way or another experiencing precarity, and for some, the more immediate threat of extinction.

With the uncertain state of planetary futures, established knowledge paradigms, models of thinking, the system of representation and habitual action are now more uncertain and provisional, insofar as they were fashioned by and addressed to an older world order. Much of life is still lived out, paradoxically, in relationship to local and regional customs, patterns of work, family ties, traditional social bonds and a solid, if misplaced, sense of the permanence and abundance of nature. But life on earth has changed dramatically, both intimately and externally, by relations and functions of global capitalist economics and sophisticated networked technologies. We are living through nothing short of a cultural revolution, a term offered by Fredric Jameson in the 1980s to account for unevenness in historical changes in the mode of production and its determinations upon social life (2002, p. 81). Media theory can no longer treat the irrevocable harm capital accumulation is wreaking on the planet, nor the extreme reactionary violence, rape, murder and genocide that proxy wars necessary to capital extraction bring about, as simply media content. The global, capitalist mode of production has to be part of a theory of networked media technology. Any account of contemporary visual culture, media and communication which ignores these realities will fail to understand how and why the networked image operates as it does.

There is a very real danger, at what is a crucial moment of the reconsideration of the politics of the image, that critical understandings will quickly become contained by institutional academic orthodoxies, turning into yet another silo in the intensified churn of commodified knowledge. The commodification of knowledge works its way through research and scholarship, such that critique is neutralised and outstripped by instrumentalised research funding demanding market-ready knowledge. This can be seen in the ways in which even progressive knowledge, which seeks to understand the image under conditions of capitalism, is parsed through its publication, much as this book and this chapter will be. Alternatively, it could also be the case, as explored in what follows, that new knowledge of image technologies might contribute to the goal of a new literacy and become more widely disseminated and incorporated into democratic and public life.

The Networked Image

The term *networked image* is primarily an abstract concept, of a provisional nature, to denote a specific computational organisation of knowledge, but it is also a pointer to and descriptor of a concrete and complex set of material operations. The networked image has an infrastructure that requires labour and capital to produce a constant energy source, the mining of raw materials, the manufacture of electronic devices, the launching of space rockets, the construction of server farms, the laying of cables, and the deployment of transmitters and receivers. On current predictions, the Internet will use 20% of the total world consumption of electricity by 2025. The global network constitutes an entire mode of production, what the neoliberal World Economic Forum has defined as the fourth industrial revolution[1] and what Jonathan Beller has termed *computational capitalism* (2018, p. 12). The network image requires the unpaid labour of billions of people who spend increasing amounts of their daily lives online and in doing so provide data for corporate platforms, what Nick Srnicek (2017) and others have termed *platform capitalism* and *the attention economy*.

Critically, Benjamin Bratton has termed planetary scale computing as 'The Stack', an accidental megastructure, which he argues is 'changing not only how governments govern, but what governance even is in the first place'. Bratton conceptualises The Stack as a complex platform, which can be modelled vertically as six layers: earth, cloud, city, address, interface and user (2015, p. 9). In the architecture of The Stack, the image can be understood to operate specifically between or as the interface and user layers. As Ingrid Hoelzl and Rémie Marie take up, a new definition of the image can be found in computational networks. In *Softimage: Towards a New Theory of the Digital Image* (2015), they argue that the video signal, which constantly refreshes the screen, removes the difference between the still and moving image and that this new temporality challenges the photographic paradigm. Importantly they recognise a central paradox in that while the algorithmic 'digital image' would appear to erode the photographic geometry of perspective and its projection, as well as the philosophy of truth that went along with it, at the level of visual perception, the photographic image continues to occupy the entire field of representation.

How to make sense of the apparent paradox of the algorithmic simulation of the photographic image was also the subject of Daniel Rubinstein and Katrina Sluis's important essay, 'The Digital Image in Photographic Culture; Algorithmic Photography and the Crisis of Representation' (2013). They describe the collision of software and image, not only as a different process of image production but also, in line with the analysis of Hoelzl and Marie, as a cultural paradigm shift, with implications for 'the ontological link between representation, memory, time and identity' (p. 25). Rubinstein and Sluis liken the paradoxical nature of the digital photographic image to the two-faced Roman god, Janus, with one side facing appearances and the other facing towards the repetition and serial reproduction of the image. They go on to say that the paradox of the networked image lies in two incompatible logics, a representational and rational system of thought, in which the image refers to the existence of something real, and a recursive and viral logic, in which the image refers only to itself. Hoelzl and Marie reinforce this view of the double face of the image, although reach a somewhat different conclusion, in amplifying the key characteristics of the algorithmic image as its programmability and operability, which they describe as the continuous actualisation of networked data. The networked image no longer has a fixed representational form but a 'signaletic temporality' capable of transfer across digital networks in real time.

Between them, Rubinstein and Sluis and Hoelzl and Marie, lay out some of the defining characteristics and behaviours of the algorithmic, networked image, as well as identifying new problems and issues of how meaning is constituted. Rubinstein and Sluis propose that undecidability is the fundamental property of the networked image, as opposed to the immaterial or indeterminate. For them the meaning of the image is no longer fixed, but rather meaning develops through the progressive accumulation of what they term a 'data shadow', which determines its visibility and currency. Meaning is ongoing in tagging, commenting and syndication. One direction for future critical analysis contained in the notion of undecidability is to pursue meaning and hence power within the algorithm, to reverse engineer software in order to reveal the agency of computation itself. This project has been taken up by Nicolas Malevé who examines the social ontology of machine vision, and is discussed in this volume (Chapter 4). Another direction for critical studies, hinted at by Rubinstein and Sluis, is one that reconsiders the paradoxical, representational face of the image, and follows the work of Giorgio Agamben, in suggesting that the image might be rethought in terms of the ways in which technologies of the network inform identity and subjectivity. Collectively they emphasise the dialogic nature of interaction in which attention should be paid, not to the appearance of the image on screen, 'but to the rhythms of repetition and recurrence, to the time signatures, to the gaps and pauses, the stuttering, the noise, the immoderate abundance that constitute the image because these are the mechanisms that make it cohere' (Rubinstein and Sluis 2013, p. 36).

Hoelzl and Marie follow a similar logic in which they see the representational image of the screen as nothing more than a lure for real processes of surveillance and control, leading them to conclude, pessimistically at the time, that the future will be defined by automated sensing systems, robotics and artificial intelligence in which humanity will be sidelined. They have since adopted a more positive account for the future human relationship with the networked image. Writing on *generalhumanity.org* in 2020, Hoelzl and Marie still argue that we need to abandon the humanist image and its pillars of visual culture but go on to say that we need to understand the image *as* 'our entanglement in a mesh of protean relations among all wordly co-dwellers'.[2] This, they propose, requires a different type of perception, an adaption of our brains, a re-distribution of the senses in which the image is understood as an *'ensemble* of shared perceptions, exchanges and thoughts across the community of the living and the non-living'. In a further text,[3] Hoelzl (2019) reiterates her understanding that the networked image is a continuous process of mutual data exchanges between human and nonhuman partakers, but now with an emphasis upon transaction as a particular form of relationality. Here she uses the term 'transactional ensemble' to characterise the 'operations taking place in digital networks', and the term 'intra-image' to define 'the transaction between the representational (on-screen) and the algorithmic (off-screen) function of the image'. The transactional image takes Hoelzl's model beyond a theory of the networked image merely as the result of the processual nature of network access, data transfer and display. For her, the image is not simply an output displaying network processes on a network terminal, it has a continuous relation with(in) the entire transactional assemblage. For Hoelzl the resolution to the paradox of the continuation of representation, or Rubinstein and Sluis's two-faced Janus, is that the representational and computational paradigms now function as a perfect synergy, because the algorithmic has evolved to operate in continuity with representation. At the level of user experience, the integration of vision and representation remains intact, and at the computational level, the image is integrated with software, making transaction possible. In cultural and media terms, Hoelzl sees the photographic image as augmented by the networked image 'with hitherto unprecedented possibilities of multi-vision, tele-vision, navigability and real-time adaptivity'. However, the caveat implicit in Hoelzl's analysis is that the changes in visuality she identifies take place in relationship to the larger system of capitalist relations.

A closely related analysis of a paradigm shift in visual culture is developed by Adrian Mackenzie and Anna Munster in their essay, 'Platform Seeing: Image Ensembles and Their Invisualities' (2019). They question how the operationalisation of visual culture can be seen, given the automation and imperceptibility of vision in computing. How is it, they ask, that images lose their stability and uniqueness yet gain greater aggregate force? As Mackenzie and Munster go on to suggest, for images to matter in contemporary computation, they must be understood as entities in a computational relationality and, in parallel with Hoelzl and Marie, see this relationality as a socio-technical image ensemble. For Rubinstein and Sluis, the image ensemble is the algorithmic image, in which the site of the visual remains located in the photographic image. For Hoelzl and Marie the image ensemble is materialised at the screen interface. Mackenzie and Munster centre the ensemble upon platform 'life' and 'perception', which they see as the conjuncture of image datasets, AI architectures, devices and hardware. It is through platforms that the image ensemble is qualitatively transformed in labelling and formatting, where the image is made platform ready. It is the image made operable that demands a reformulation of visuality, what Mackenzie and Munster term 'platform seeing', a mode of perception they define as 'invisual' because it is no longer qualitatively of the order of representation. For the purposes of developing a politics of the networked image, invisuality is both a new description of image operativity, as well as the identification of its problematic within the social relations of the reproduction of visuality. Mackenzie and Munster articulate the nonrepresentational mode as the production of a 'present image', not observable by the seeing subject, as direct mirror or reflection, but as a new model of truth. The 'reality' of the present image lies in machine learning systems held, diagrammatically, through the reflowing relations of the image ensemble in which materialities and experiences are generated.

In *Signal. Image. Architecture* (2019), John May develops an equally important analysis of the status of the contemporary image based on computation. He does this from the perspective of architectural practice in considering the historic medium specific differences between drawing, photography and imaging. In doing this, he also establishes three philosophic axioms, drawing upon the philosophic work of Vilém Flusser and Bernard Stiegler, which state,

> i) there are no pre-technical thoughts; ii) nothing technical is ever merely technical; and iii) the specific conception of time embedded in a technical system is inseparable from the forms of thought and imagination that system makes possible or impossible.
>
> *May 2019, p. 36–39*

May starts from the perspective of a chronic confusion in architectural practice, arising from a lack of interest in defining the differences between drawing, photography and image. Following Walter Benjamin, May distinguishes between drawing, as a static hand-mechanical depiction, photography as a chemico-mechanical visual format, and 'digital' imaging, based upon a process of detecting energy emitted by an environment, translated into discrete, measurable electrical signals. From here May is able to account for what Mackenzie and Munster term the 'present image' in similar terms as, 'the outputs of energetic processes defined by signalization, and these signals, in their accumulation, are what we mean when we say the word data. Images are data, and all imaging is, knowingly or not, an act of data processing'.[4] May stresses that images are a telematic technical format, always already mathematically quantified, whereas the photographic image amounted to an historic regression of visual mathematisation. May concludes that photographs and images have virtually nothing in common, that they are two incompatible technical formats, belonging to two competing epistemic visions of the world. For what is being followed here, signalisation changes everything: linear historical time, materialised in photographs, texts and drawings, is superseded; by image models signalised in real time, in which all possible futures at theoretically present. As May says,

> Real time is the time of statistical thought, in which futures knowable and unknowable are posed simultaneously, some more calculably probable than others, but all possible. This probabilistic conception of time is fundamentally different from the linear, mechanical conception that structured orthography.
>
> *2019, p. 36*

There is a great deal of synergy between the texts being considered. There is, for example, a general agreement that the technical architecture of the networked image, developed over the past three decades, now produces the image as a relational ensemble of humans and nonhumans, networked in real time, through the operations of data. It is upon the image understood as algorithmic, transactional, platformed and signalised that a radical paradigm shift in established ways of seeing is recognised, a cultural change in which the image of thought itself is fundamentally reconfigured. In recognising such a paradigm shift, an important claim is made; that the new form of visuality is based upon a system which itself cannot be seen. Expressed in Mackenzie and Munster's term of the invisual, it is impossible for the observing subject to see the operability in any holistic or meta-observational display, with 'observation operating in and through the image but not of the order of the visual'. May also notes that the processes within the topology of electrical signalisation are concealed from perception by their size and speed. The paradox of this moment of the visual in culture is that the networked image operates in two simultaneously perceptual/cognitive registers, one propping up an established code of visuality, the other busily undermining it. The observable screen image and the signalised image are coupled in the algorithmic image but have different destinations. The screen image drives towards satisfying the socius and with its desire and the unconscious, while the signalised image drives towards optimal financialisaton, capital and exchange. This is another way of putting Mackenzie and Munster's point that visual technologies and practices continue to expand what can be visualised, even as the visual itself is being evacuated. The interface between the domain of insensible data and the domain of the screen image, which is sensible to everyday life, is the programmable image. This is what Hoelzl has termed the synthesis between the algorithm and the codes of representation, or what May has termed *pseudo-orthography*, a simulation which produces the image model. At this point in assessing the networked image, the question that comes to the fore is, if visuality has left, or is leaving, the image, what is left for representation to do?

Bridging Worlds

The Internet is always already on; it is the computational substrate of a global economic infrastructure and telecommunication. Being online, or online being, is experienced individually as coterminous with real time and space, such that everyday life is now lived in the virtual of real-time computing. This coupling of networks and everyday activity raises a host of new problems and questions for the practices of everyday life and our understanding of how life has changed. In Web 1.0 terms, the object being framed here could still be grasped as media, but the greater imbrication of networks in the everyday life world has reformatted and dissolved media distinction. In attempting to reveal a politics of media and networks, attention to the relationship between the technicity of the operations of the networked image and the mediatic cultural forms of the online becomes crucial. How are the new relationships between the nonvisual or 'invisual' operations of data, and the everyday subjectivities produced online by the persistence of representational images, to be conceptualised? So far, the networked image is being conceived as, the modus operandi, the default of the ensemble of computational and human transactions in the operations of data signalisation and platform processing. However, as already noted, such operations conceal the agency of their own logic. May's analysis leads him to declare that the process in which 'the visible, physical, and material dimensions of collective life were ceding political primacy to its invisible, electrical, topological, and ecological aspects' is complete and that while the politics of urban life must be enacted, it would be a mistake, 'to see this everyday politics as the primary terrain of political theorization'. If by this we are to understand that power is now conducted in the invisible realm of data, what then of the politics of everyday life, in which the screen and image partake? Without counter technical efforts to reveal or unstitch the processual coding and aggregation of data, the network offers no reflexive, intra-relational position from which its own image can be viewed, and at the same time the politics of everyday life have been evacuated. This is what has been termed for some time as the crisis of social democracy and its representational form.

What evidentially lies in the wings of the analysis of the techno-image ensemble, waiting to make an appearance, is the question of the networked image's relationship to economic and political forces, to the logic of the entire mode of global production. The planetary scale of computational networks requires a planetary politics of the networked image, which is also to say a politics of the planet, at one and, at the same time, a politics of the organic and inorganic, the human and nonhuman, of the body, technology and history. It will be a reflexive politics of epistemological mediation, of making sense of sense itself, and then considering how to act upon such sensing. For a long time now, it has been the norm to say politics affects media and that media have political affects, but this formulation belongs to earlier foundational theories of the binary relationship of subject and object, in which mediation was posited as standing between the external world and the human perception of it. Contemporary critical cultural theories, drawing upon non-foundational currents in phenomenology, process philosophy, pragmatism, science and technology studies, and feminist critiques of techno-science, have challenged the binaries of positivist scientific models, in seeing both object and subject in relational and processual terms. As has already been noted, the image ensemble has been framed by drawing upon such work, precisely because to be human is to be imbricated in machinic networks. But what has also come back into critical view, given the network's relation to financialisaton, is the application of Marx's analysis of capital and labour, in accounting for subjectivity in terms of the network's relationship in and to the social relations of the reproduction of the global mode of capitalist production.

Both Marxist and companionable non-foundational critical theory connect the image in visual culture to the environmental planetary and human crisis, which manifests itself daily on a global scale and figures daily in 'media' discourse. There is a greater recognition of what Jonathan Beller calls 'the global warming of the material substrate of our thought', in which these 'hot materialities' signal a 'return of the repressed of centuries of idealism, alienated science and all its attendant if unfathomable colonial violence' (p. 76). Beller convincingly recasts the historical development of capitalist wage labour, private property and exchange value as the development of computational capitalism in which information is the dominant form of alienation of humans from nature.

For the problem being pursued here, of the affects, upon subjectivity and everyday life, of the visible/invisible operations of the networked image, Beller offers two important arguments. Firstly, in his 2016 essay, 'Informatic Labor in the Age of Computational Capital,' he argues that as a result of the informationalisation of social practice, culture is financialised. This means that culture is no longer separate from production, if indeed it ever entirely was, but more to the point, culture, and here we can say the image, manages information flows from user to capital, in what Beller formulates as Image-Code-Financialisation. Recognising the invisibility of machine operations, Beller argues that the screen/image functions as a valorisation process of capital, while concealing the greater crisis of control of the management of the built environment and workplace. Beller is claiming that what was previously valued as the cultural life of imagination, interiority and the self has been commodified by information and that this has had a profound effect upon what is understood as wage labour.

While people across the globe still work in factories and on assembly lines, the key commodities they produce are no longer objects with a singular point of sale but rather 'tethered to computable information and anchored to a distributed material system with multiple points of interface'. In Beller's, terms, the wage labourer has been structurally reduced to a function, a biochip in a ubiquitous computational armature (p. 163) computation has created unpaid labour in the attention economy. This leads to the second instructive point for considering a politics of the network image from an earlier essay, 'The Cinematic Mode of Production: Attention Economy and the Society of the Spectacle' (2006), in which Beller elaborates the idea that to be a spectator is to perform value-productive labour. Beller defines cinema and, by extension, television, video and the Internet as deterritorialised factories in which we work, and for this reason, he says 'cinema brings the industrial revolution to the eye'. In what he sees as a totalitarian social space, attention, 'in all forms imaginable and yet to be imagined (from assembly-line work to spectatorship to Internet-working and beyond), is that necessary cybernetic relation to the socius – the totality of the social – for the production of value for late capital' (p. 92). Beller's formulation leaves the image, as a productive means for the transfer of attentional biopower to capital, which sustains and perpetuates all forms of subjugation. The argument that culture and the image are an inseparable part of labour, is one answer to what the residual representational image is doing, now that visuality and its coding of power have ceded to data. In real political terms, the image as the vehicle for the transfer of biopower to capital raises important questions about the naturalisation of differential labour power, as well as, it might be added, problems with a totalising theory of labour. Crises, contradictions and breakdowns are always in evidence in social relations and do more than simply ensuring the reproduction of the global mode of production. It is at the extremes of capitalist excess that other possible futures can be imagined and acted upon.

The Smartphone

The complexity of material and subjective relations between technological intimacy and geo-political forces is nowhere made more manifest than in the smartphone of which the exemplar is the Apple iPhone. In 2020 there were approximately 5.27 billion unique mobile phone users in the world, with a predicted growth rate of 2.3% per year. It is now almost trite to say this, but the smartphone is never far from the body of its user and the amount of user time spent 'checking' smartphones ranges from 3 to 6 hours per day. The figures are almost meaningless, other than to underline the fact that life has changed. Apple's net worth in 2021 stood at $2.43 trillion, exceeding the GDP of countries like Italy, Brazil, Canada and Russia and would take the place of the eighth richest country in the world. The company's service sector has a profit margin of 60% and iPhone products a margin of 35%. Half of the world's iPhones are made in a sprawling factory complex in Zhengzhou, under license to a Korean-based company Foxconn. The Zhengzhou factory employs as many as 35,000 people in what is known locally as iPhone City. In the months preceding the release of a new iPhone, the factory can produce 500,000 phones a day, or up to 350 per minute. The workers live in dorms in 12 story buildings around the perimeters of the factory. Foxconn's iPhone factory in Zhengzhou has an assembly line which involves 400 individual operations, with the worker repeating one task on a 12-hour shift, typically handling 1,700 iPhones.[5] The average wage of an Apple shopfloor worker in the Foxconn factories in China is $1.62 per hour, less than half the recommended living wage of about $3.77 per hour.[6] Such facts, reported in Europe and North America, are intended to underline the poor conditions of workers in the electronic sweatshops of the world and in particular reinforce a Westernised view of China's apparent disregard for human rights. Smartphones are also manufactured in India, Korea and Vietnam and component parts are also made in Malaysia, Czech Republic and Thailand.

The rare earth metals used in smartphone manufacture are mined from Peru and Chile to Mongolia and Africa. The mining of coltan serves as a further example of the normalisation of the differential exploitation of global labour involved in maintaining the network. In the first decade of the 20th century, approximately a quarter of the tantalum ore, used in smartphone micro-capacitors, came from the Democratic Republic of Congo, a country in which three quarters of its population live below the poverty line. The ores lie close to the surface in the North Kivu province, bordering Rwanda and Uganda, and are extracted by coercive labour of artisanal miners, men, women and children who dig it out of the ground with picks, shovels and bare hands. Coltan mining in the Congo has been categorised as a conflict mineral, because of its political and military links to human rights abuses, illegal profiteering and civil war. In 2010 the US Congress passed an act requiring companies using conflict minerals to disclose their sources, which has been criticised as having a worsening effect on the conditions of Congolese artisanal miners, because it interrupted the 'illegal' status of the mining, putting them out of work.

The example of the global assembly of smartphones is no exception to the global mode of production of all commodities, based upon geo-political extraction and manufacture, organised upon driving labour costs down and profits up, the logistics of which are produced by computational calculation. As Sean Cubitt (2006) has noted, '[a]ll those minerals and sub-assemblies in our computers and phones are nothing but costs until they have been sold: the faster they are converted back into money, the better'. Differentiated and divided labour is necessary to capital, including a reserve army of labour, reproduced globally in social gulags and in relation to the immediate and future needs of capital. Smartphones require, organise and reproduce global differentiated labour power, which is to say the social relations of the reproduction of divided, exploitative and unequal labour power necessary to their production. As Jonathan Beller puts it, 'The iPhone is a particularly good example, because even as the A-side of its screen is immersed in networks and clouds, the B-side depends on a network of labour practices that are effectively forms of enslavement' (2018, p. 158). The smartphone hardwires alienated labour. As Susan Ferguson (2017) elaborates her account of social reproduction:

> What capitalists truly need and want is not labour or labourers, but *alienated labour*. At the same time, the fact that they can't get this without its human packaging is their biggest problem. It means that, always and everywhere, the bodies and minds of workers can and do push back against the dehumanising dynamic they are part of.

It is then a supreme irony, that at the launch event for the latest iPhone 13, Tim Cook, Apple's CEO, announced the phone as 'your new superpower'. The iPhone 13, cinematically imaged, on your laptop or smartphone, encodes the egalitarian myth of freedom, creativity and ambition made possible by advanced technology, a Californian and US dream, the ultimate Silicon Valley techno-imaginary.[7]

What Remains of the Image

The reproduction of differentiated global labour power, necessary to produce smartphones, directs attention to the processes by which the network reproduces globally unequal conditions through the social relations of its own production. To be in networks, as discussed earlier, is to be in the social relations of the reproduction of the computational mode of production. Thus, to be in networks is to participate in the reproduction of labour power, not only one's own, necessary to the functioning of networks, but the reproduction of globally unequal differentiated labour of the production of the network itself. And as the argument laid out here has attempted to evidence, the processes of reproduction are conducted by network operations and interaction, of which the image is its paradoxical interface. The image conceals the extractive labour of network operations within itself, whilst the image simulates a representational reality that normalises the inequalities in the conditions of labour. Representation reinforces the 'sensible' interpretation of objects and events and occludes the solidarity of labourers. A point echoed by Hito Steyerl (2014), in saying that representation stands to the present condition, as analogue photography does to the networked photograph, working with probabilities and betting on inertia, with the consequence of making unforeseen things more difficult to see. As she says, 'The noise will increase and random interpretation too. We might think that the phone sees what we want, but actually we will see what the phone thinks it knows about us'.[8]

Representation, now as the screen image, constitutes the continuity of the legible symbolic logic of visuality, even while visuality functions as a derivative in the operations of the network. This leads us to ask, where does the image stand in network social reproduction, and in effect, what is the image's relationship to what is still taken as media? In considering photography as a representational mechanism, Daniel Rubinstein (2020) observes that representation is a conservative force. Representation itself doesn't change, even though the content of an image does, creating an impression of stability and continuity. As he says, 'Under the rule of representation it is impossible to conceive the prospect of genuine revisionism, of a situation where representation, as a mechanism that has subjectivity at its core, is replaced by another way of knowing and thinking'. This led Rubinstein to conclude that by accepting representation, 'we make it impossible to imagine the potential of a human being who is not experiencing the world as a subject, for whom the world in not an image or a picture' (p. 338).

Political Theory

Here there is a need to pause to consider the relationship between representation and power. The hegemonic formation of representation has long been part of a socialist critique of capitalist media, drawing explicitly upon cultural Marxism. But within this critical tradition, it has also been recognised that ideology critique has political and ontological limits. Politically, in its unreflexive acceptance of the totalising structures upon which the function of ideology rests and the ultimate passivity accorded to the individual subject. Ontologically, the limits of ideology critique rest upon the theorisation of the fixed externalised relations of subject and object in which objects have no agential affects upon subjects. Readers may be familiar, by degrees, with the parting of the ways of French intellectual Marxism from Communist Party alignment, following May 1968, which continued until the eventual collapse of the USSR. This marked not only a final break with Stalinism and doctrinaire and reductionist interpretations of historical materialism but also a more fundamental break with the philosophical underpinnings in Enlightenment scientific thought upon which Marxism also rested. The subsequent story of left intellectual critique is, of course, traceable through the development of post-structuralist and postmodern critical theory. Contemporary critical theory is, however, still grappling with the question of the nature of the unequal distribution of power. In wanting to understand how power can be articulated in networked organisation, how representation continues to profit from data and the labour power which produces it, continues to demand close attention. Connecting the social relations of the reproduction of labour and the collective interest of workers in gaining control over their own labour power bears upon the operations of networks in which representation is entailed.

Fredric Jameson revised an historical materialist theory of the interpretation of cultural artefacts in his book, *The Political Unconscious* (2002), in which he proposed that culture functions not only as instrumental ideology but also as the utopian promise of collective freedom discoverable through a positive hermeneutics in all cultural artefacts. Jameson formulates the task in the following terms, 'The assertion of a political unconscious proposes that we undertake just such a final analysis and explore the multiple paths that lead to the unmasking of cultural artifacts as socially symbolic acts' (2002, p. 5). At the beginning of his exegesis, Jameson quotes at length from Gilles Deleuze and Felix Guattari's, *Anti-Oedipus*, on the ontology of the unconscious, in which they are at pains to point out that the unconscious represents nothing, but it nevertheless produces, and that as the unconscious has no meaning only problems of use, desire makes its entry, not in the form of transcendence, but in terms of immanence (2002, p. 6). Jameson takes Deleuze and Guattari's retheorisation of desire as a demand for a new immanent, anti-transcendent materialist model of interpretation. In 1983 Jameson saw that a Marxist analysis of culture not only needed to find a way of retaining a negative hermeneutic to demystify ideology but to open up the utopian potential of the political unconscious, 'as the symbolic affirmation of a specific historical and class form of collective unity' (2002, p. 291). Deleuze and Guattari's work is in large part a rejection of the reductivism in orthodox Marxism's interpretation of a theory of ideology, on the grounds that political economy takes no account of the libidinal economy as a production of social relations. However, Benoît Dillet (2016) revisits ideology in claiming that Deleuze and Guattari's 'Noology', in which they included the image within their studies of the system of thought, was consistent with a general theory of ideology. For Deleuze and Guattari, expressions and statements intervene directly in productivity as meaning or sign-value and hence could account for images that create adhesion to the system of capitalism. How then, finally, to account for the networked image's relationship to power and the persistence of a representational psycho-social semiosis, when both the use and exchange value of the image lie in its data operations?

Time Again

Jonathan Beller argues that there is no longer any distinction between media and computational capitalism because media is a fully incorporated function of capital extraction. From a related perspective, Lipovetsky (2005) argues that the replacement of linear historical time by multiple temporalities has led to a situation in which the distinction between contemporary media and heritage no longer holds. From both perspectives, it can be seen that we live and labour in media. All specific media cultural forms, which owe their origins to the mechanical analogue, are simulated by the programmable image, as the interface within the flow of data. And yet, because of computation's need for representation, media is still experienced as photography, film and video in the cultural forms of cinema, television, print and their online doubles. We can say that at the visible, cultural interface of the networked image, media is now primarily and paradoxically experienced as a nostalgic loop. Nostalgia is produced through the reproduction of the representational code, but without reference to the historical mode of its own production. John May underlines such a view, arguing that whilst the medium of orthographic drawing linked people in historical time, through the image and its archive, now the image delivers only infinitely revisable versions of its data. Hence, in May's argument, any version of the billions of images deliverable by data is equally possible. Hoelzl and Marie make a parallel argument about the networked image's relation to time, which is no longer historical time, but rather the image-screen as a local access to and visible part of data exchange, in which the image is defined in terms of the speed of data access and transfer. Hito Steyerl also points out that the life span, spread and potential of an image in social media are defined by participation. From the revised perspective of the shift from representation to participation, perception to performance, and from the ocular-centric to the statistical, what does the repetitive sequencing of representational images across screens do and signify? What forms of attention are engaged by the perpetual supply of photographs, videos, music, films, games, texts and sound files, relayed by Facebook, Apple, Amazon, Netflix and Google and their subsidiaries?

Trans-hypermedia

Transmedia analysis suggests that to understand the meaning of any single media product it is necessary to account for its place in an extended commercial network of related content. In media studies, Web 2.0 transmedia analysis usefully drew attention to the extended narratives associated with media marketing. It adopted a polysemic approach to images, characters, storylines, scenes and plots, which could be tracked as they were reproduced in different media contexts. In this sense, transmedia analysis is suggestive of the ways in which images selectively travel, morph and aggregate through distribution and circulation. The idea of transmedia, like that of the networked image, is a useful abstraction for the state of remediated representational media. The convergence of media production in the hypermedia environment of software expands the idea of transmedia to include not only the crossing of image reception and meaning from one media to another but also now with the polysemy of the material constitution of the programmable image. What started out in the 1990s as multimedia has exceeded the boundary of combinations of discrete media and now exists as continuous programmable media through software production, leading to a state of what might somewhat awkwardly termed *trans-hypermedia.* Such a term would doubly frame the fluid, seamless interchange of graphical vectors producing a programmable image and cultural forms in which it is produced. The differences between moving and still, durational and static, photographic and ortho-graphic, animated or live-action, and two or three dimensional are overridden or overwritten by software and signalling.

Trans-hypermedia defines the technical and cultural translation of one image into another medium, and in the case of the networked image, it is relayed as both image-data and data-signal. With the programmable image and signalisation, the user even more so produces and is produced by the network. Obviously, this not to say that the global media entertainment complex has ceased production in the face of user generated content, far from it, after all this is media as capital extraction. However, the algorithmic logic of corporate image production depends on users through networked operations in which consumers create meaning through their choices. The tracking and profiling of users and their media preferences combine with commercial production decisions and cycles in a recursive cycle of liking and sharing. Trans-hypermedia can be understood as the vehicle of Jameson's symbolic exchange, which is both individualised and aggregated in commodity form. To translate such aggregates would get nearer to articulating Jameson's political unconscious or nearer understanding a Deleuzian desiring machine. It is the user who enters the libidinal field of desire in consuming images and whose choices aggregate in media production in the recirculation of ideas, expressions and narratives. The aggregation and distribution of trans-hypermedia exist on a global scale and this makes (all)media transcultural, although unequal in translation, in which embedded cultural historical knowledge, long ripped from the fabric of tradition as, Walter Benjamin crucially noted, can be understood as a libidinal flow which has no other time but that of the present.

Closing

More could and needs to be said, but this chapter must close by returning to the three ways in which it proposed the networked image might be understood. Firstly, it considered the network as a relational assemble of human and nonhuman actors in which the image functions as an interface to the operations of data. Secondly, the network was considered a substrate of computational capitalism in which labour power was reproduced by the financialisation of data. Finally, the programmable image was defined as a new condition of the global aggregation and circulation of images as trans-hypermedia. There are problems and limits in each account, but what comes through is the fundamental scale of the paradigm shift in the human relation to the world entailed by the networked image. What needs more thought is how to build a complete and accessible language of the image in computational capitalism. How can a new discipline of knowledge of the visual be constructed to traverse the dimensions, the peaks and troughs of the global, algorithmic and telecommunication landscape, which includes its infrastructure, operations, interfaces, tools, hybrid images and user behaviours? Such a project in the future will be able to articulate the reality the networked image produces and is productive of. The call for new forms of computational literacy is in the final analysis, a form of political literacy.

Notes

1 World Economic Forum. www.weforum.org/focus/fourth-industrial-revolution.
2 Hoelzl, Ingrid, and Rémi Marie. 2020. 'No Image No Cry'. *GENERAL HUMANITY* (blog). 1 January 2020. https://generalhumanity.org/2020/01/01/no-image-no-cry-by-ingrid-hoelzl/..
3 Hoelzl, Ingrid. 2019. 'Image-transaction'. *Multitudes* 77 (4): 129–40. https://doi.org/10.3917/mult.077.0129.
4 Adrian Mackenzie and Anna Munster (2019) Platform Seeing: Image Ensembles and Their Invisualities. https://journals.sagepub.com/doi/full/10.1177/0263276419847508.
5 Harrison Jacobs (2018) Inside 'iPhone City,' the Massive Chinese Factory Town Where Half of the World's iPhones Are Produced. www.businessinsider.com/apple-iphone-factory-foxconn-china-photos-tour-2018-5?r=US&IR=T. May 7, 2018, 5:5.
6 Apple Soars, Workers Suffer. www.greenamerica.org/end-smartphone-sweatshops/tell-samsung/apple-soars-workers-suffer.
7 See, for example the promotional video for Apple iPhone 13. www.apple.com/uk/apple-events/september-2021/.
8 Steyerl, Hito. 2014. 'Politics of Post-Representation'. Interview by Marvin Jordan. *DIS Magazine* http://dismagazine.com/disillusioned-2/62143/hito-steyerl-politics-of-post-representation/.

Bibliography

Baudrillard, Jean. 2012. *The Ecstasy of Communication*. Los Angeles, CA. Semiotext(e).
Beller, Jonathan. 2006. *The Cinematic Mode of Production: Attention Economy and the Society of the Spectacle*. Hanover, NH: Dartmouth University Press.
Beller, Jonathan. 2016. "Informatic Labor in the Age of Computational Capital". Lateral Issue 5.1. Chicago. *Journal of Cultural Studies Association*. https://csalateral.org/issue/5-1/informatic-labor-computational-capital-beller/
Beller, Jonathan. 2018. *The Message Is Murder: Substrates of Computational Capitalism*. London: Pluto Press.
Benjamin, Walter. 2008. *The Work of Art in the Age of Mechanical Reproduction*. London: Penguin.
Bratton, Benjamin. 2015. *The Stack: On Software and Sovereignty*. Cambridge, MA: MIT Press.
Bolter, Jay David and Richard Gruisin (2000) *Remediation*. Cambridge, MA: MIT Press.
Cubbit, Sean. 2016. *Against Connectivity*. Paper presented at in/between: cultures of connectivity: NECS European Network for Cinema and Media Studies 2016 Conference: Potsdam, Germany, 28–30 July 2016.
Dillet, Benoit. 2016. "Deleuze's Transformation of the Project of Ideology Critique: Noology Critique." In *Deleuze and the Passions*, edited by Ceciel Meiborg and Sjoerd van Tuinen, New York: Punctum Books, pp 125–146. https://doi.org/10.21983/P3.0161.1.00
Ferguson, Susan. 2017. 'Social Reproduction Theory: What's the Big Idea?'. *Pluto Press*. 23 November 2017. https://www.plutobooks.com/blog/social-reproduction-theory-ferguson/.
Ferguson, Susan. 2019. *Women and Work: Feminism, Labour, and Social Reproduction (Mapping Social Reproduction Theory)*. London: Pluto Press.

Fuller, Matthew and Eyal Weizman. 2021. *Investigative Aesthetics*. London.Verso.

Gilles Deleuze and Felix Guattari. 2004. *A Thousand Plateaus: Capitalism and Schizophrenia*. London: Continuum Press.

Gilles Lipovetsky. 2005. *Hypermodern Times*. New York: Polity Press.

Hoelzl, Ingrid and Rémi Marie. 2015. *Softimage: Towards a New Theory of the Digital Image*. Bristol: Intellect.

Jameson, Frederic. 2002. *The Political Unconscious*. 2nd ed. London: Routledge.

Latour, Bruno. 2018. *Down to Earth: Politics in the New Climatic Regime*. London: Polity Press.

Mackenzie, Adrian and Anna Munster. 2019. *Platform Seeing: Image Ensembles and Their Invisualities*. https://doi.org/10.1177/0263276419847508.

Manovich, Lev. 2001. *The Language of New Media*. Cambridge, MA: MIT Press.

May, John. 2019. *Signal Image Architecture*. New York: Columbia Books.

Rubinstein, Daniel and Katrina Sluis (2013) 'The Digital Image in Photographic Culture; Algorithmic Photography and the Crisis of Representation'. In *The Photographic Image in Digital Culture*, edited by Martin Lister, 2nd ed. London: Routledge.

Rubinstein, Daniel. 2021. "Fractal Photography and the Politics of Invisibility." In *The Routledge Guide to Photographic Theory*, edited by Mark Durden and Jane Tormey. London: Routledge.

Srnicek, Nick. 2017. *Platform Capitalism (Theory Redux)*. London: Polity Press.

Steyerl, Hito. 2013. *Too Much World: Is the Internet Dead?* e-flux Journal 49 (November 2013). www.e-flux.com/journal/49/60004/too-much-world-is-the-internet-dead/.

2

THE NETWORKED IMAGE AFTER WEB 2.0

Flickr and the 'Real-World' Photography of the Dataset

Katrina Sluis

Networked Image

In 2008, Daniel Rubinstein and I began a series of articles which sought to inscribe the affordances of socio-technical infrastructure into photographic scholarship (Rubinstein and Sluis 2008, 2013a, 2013b). We observed that canonical discourses of photography had either fetishised the loss of the real under the 1990s paradigm of post-photography or dismissed the digital image as a spectral and immaterial surrogate unable to compete with the analogue print. Writing in the aftermath of the convergence of the camera, phone and Internet, we proposed the 'networked image' as a paradigm or condition of the authorless, contextless screen-based image, borne of the massification of the snapshot. Whilst photography has always been portable, mobile, and valued as a reproductive technology, the networked image signalled a post-representational turn in which the iconicity of the image was being hollowed out and declining as a site of cultural and socio-economic value, while the relations between images became more valuable.

Through practices of image sharing, the networked image's agency was materially bound to database architectures of an increasingly centralised Web 2.0, in which the photographic image was positioned both as a cultural residue and software output shaped by human and nonhuman relations that lie behind the screen – increasingly inaccessible to scholars of photography. When materialised in the web browser, the networked image adopts the comforting rhetoric and reassuring continuity of photographic representation, whilst generating new topologies via its computational back end. Under conditions of ubiquity, individual images proliferate yet struggle to be seen, whilst the metadata generated by sharing and social annotation ensure they became legible to machines. We called for a re-thinking of the image as processual, algorithmic, softwareised – a product of algorithms, not just in terms of its production but also its circulation. In 2013 we concluded: 'the

DOI: 10.4324/9781003095019-4

image within the network is doing something other than showing us pictures, and it is doubtful if we have the right vocabulary to address this image economy' (Rubinstein and Sluis 2013a, p. 156).

Today the networked image – the fugitive, repetitive, 'ocular white noise' produced by image circulation – has become the dominant paradigm of photographic culture. Of the estimated 1.4 trillion photos taken in 2020, 90% of these were on mobile phones (Carrington 2020). The range of cultural forms which the networked image cloaks itself in – screenshot, snapshot, meme, selfie, cctv feed, tinder headshot – continues to creatively expand, whilst image sharing remains technically fused to processes of platformisation, the epistemologies of search and the logistics of data warehousing. At the photo-sharing interface, companies celebrate and exploit the representational regimes and genres of twentieth-century photography, whilst at the back end, subject–object relations are eclipsed by pattern-randomness, database-interface, query-result and structured by economic imperatives. It is this situation and the circularity it involves in producing the networked image which this chapter addresses.

In the past two decades the exponential expansion of image circulation facilitated by photo sharing has been paralleled by a turn in the field of computer vision towards data-centric approaches which rely on the voracious collection of ever greater constellations of images. These two developments are not disconnected: millions of snapshots circulated by image-sharing communities have become the basis on which modern machine learning systems recognise and categorise the world. Social media images are harvested, cleaned, labelled, repackaged and re-circulated by computer scientists at scale as image datasets, training data used to improve the accuracy of machine vision. In terms of image sharing, this has created a somewhat circular situation. On the one hand, image ubiquity demands the automated systems of classification to determine the value and relevance of photographs which circulate beyond the limits of human attention. On the other hand, the classificatory powers of machine learning depend on the unfettered circulation of images and the creative labour of the users of photo-sharing platforms.[1] As Nicolas Malevé (2021, p. 35) suggests, through this process, the contemporary dataset functions as a 'large cache for transient networked images' and, as I argue, a site through which the cultural and technical imaginaries of the networked image converge.

The favouring of ever-greater models for machine intelligence, which depend on the harvesting of user data, has come under significant criticism for its perpetuation of racist and sexist stereotypes through problematic systems of classification and data curation, as well as the environmental costs of training such models[2]. An important body of work has engaged with image datasets from the position of the annotator who verifies and labels images, and the social ontologies of the computer lab, which support this process (Malevé, this volume, 2021; Miceli et al. 2020). Less attention has been given to the ways in which the personal snapshot has been mobilised in the photographic pipeline of machine vision to meet the need for 'real-world' photography as a proxy for human visual experience. In what follows I return to the context of the networked image, once more with Flickr in mind, to discuss the significance of photo-sharing infrastructure in shaping contemporary AI visuality. The cultural form of the amateur snapshot is centrally entailed in the problematic of contemporary visuality because not only does it continue as the popular and ubiquitous form of image-making, but importantly the snapshot also feeds the epistemologies of computer sciences as training data. Whilst the URL flickr.com points to an ailing platform and community, the abandoned image feeds its users have long escaped, reanimated and given new prominence in machine learning datasets, oblivious to their original authors. In the process, computer vision researchers have become unlikely agents of image circulation through their enduring dissemination of subsets of Flickr, now hosted on university and corporate websites. By tracing the flight of the amateur photographer from Web 2.0 mythology to contemporary AI industries, this chapter situates the networked image as a site and set of relations between back-end database and cultural interface, where modes of representation and photographic practices compete, are cannibalised and become reinscribed in new systems for image evaluation, classification, sentiment analysis, locative media, biometric systems of facial recognition, automatic photo annotation, organisation and curation. In mapping the distance between Flickr as community and Flickr dataset, the networked image is therefore positioned as an interface which translates between the representational currency of the photograph and the operations of the database.

Circulation

While Western society contends with the socio-political machinations of Facebook, Instagram and Tik Tok, the survival of Flickr – now a Web 2.0 antique – reveals the challenges of sustaining photo-sharing communities during a period where monetisation strategies have turned to economies of data extraction based upon images. From the position of the pandemic 2020s, the web has dramatically changed in the years since Flickr appeared in web browsers, with a youthful spring in its step, in perpetual beta form. Babies who were born on Flickr are now teenagers, whilst my own Flickr photostream remains frozen in 2009. With Flickr's waning, an older Web 2.0 vocabulary has declined in mainstream technology discourse, with all its participatory rhetoric: discussions of folksonomy, life logging, geotagging, mashups and empowered citizen curators have been relegated to dark corners of museum tech conferences and media studies seminars. The myth of the creative amateur as a producer of user-generated content has been morphed into the dream of the financially empowered influencer. A Google search on 'the sharing economy' will now point you to the dark arts of Airbnb, Uber, TaskRabbit and Fiverr rather than the open knowledge ecosystems of Wikipedia. Today, as Flickr approaches its 16th year of existence, it is variously described as a foundational piece of the web and a faded ghost town, struggling to find its place in a crowded photographic marketplace.

A quick glance at YouTube and it would appear I am not alone in remembering Flickr. There is an emerging subgenre of videos produced by a collective of grieving photographers, who lament Flickr's decline and its adoption of an increasingly expensive subscription model to fund its infrastructure.[3] In these missives, a despondent community of image makers unites in their collective grief about what the web has become, offering criticism of an overpopulated and unimaginative photographic marketplace. As one commenter cries: 'We as a photography community need somewhere to share our work in its full uncompressed glory!' Others confess that they have strayed to other photo-sharing platforms but remain unsatisfied: rival platform 500px is overrun by bots offering kind of emojis and empty platitudes, while on Instagram 'the combination of baffling rules regarding content, censorship, algorithms, promotions just makes it a black box'. Positioned against these competitors, there is a longing for a space for building an 'authentic' and 'true' photo-sharing community in an age of celebrity likes and swipes. One cannot help but share the exasperation of one YouTube user who asks: 'How is it that the Internet might be less useful now than it was more than ten years ago?'

This question begs both a political and technical answer which goes some way towards understanding how in networks the desire for and increased possibilities of making images contain the contradictory forces of cultural use value and capital exchange. The planetary-scale circulation of images has produced an increasingly automated field of visual production which challenges not only the limits of human attention and perception but also the sustainability of online communities. Flickr founder Caterina Fake famously described how she and colleague George Oates were able to personally greet the first 10,000 signups to the platform. Five years later Flickr's infrastructure was sustaining 3 billion photos served at 40,000 per second, using 6 petabytes of storage, hosted on 62 databases across 124 servers, with about 800,000 user accounts per pair of servers (Allspaw and Hammond 2009).

The parallel growth of user experience design and algorithmic intermediaries would appear to suggest that whole fields have been devoted to making the scale of online cultural production more navigable and user-friendly. Interfaces are more immersive, responsive and seductive. Advances in digital marketing and advertising technology seek to deliver only the most relevant and hyper-targeted products to social media feeds, and an array of platforms promise to automagically organise photos, friends and social lives. In answering the need to help users navigate the photographic glut and restore navigability, computer vision research has also stepped up. At the photo-sharing interface, AI assistants promise to take the pain out of sharing, helping photographers to effortlessly curate their best work for their distributed audiences. VSCO, a mobile photography app, promises an AI, Ava – who can look at photographs 'just like a human', whilst EyeEm's Eyevision algorithm promises to surface only the most relevant and thrilling images from the millions of images it processes each day. At the same time, an older web discourse promoting an open ecosystem of knowledge, facilitated by the centrality of the 'user', the utility of the hyperlink and peer-to-peer distribution has largely been eclipsed by the platformisation of the web, and its parallel colonisation by ad tech which generates value by keeping human bodies immersed in an endless scroll. Developers of photo-sharing platforms must 'surface beautiful images at scale' for the benefit of their users, whilst using the output of the community to produce the algorithmic models through which photographic ubiquity might be curated.[4]

Photo-Sharing Communities

Flickr's innovation was in its embrace of the digital snapshot – not as a stepping stone to a print-on-demand photo album, but as a significant and worthwhile site of ludic pleasure. In this respect, it departed from the model promoted by companies Ofoto, Shutterfly and Snapfish, the websites of which operated as cold storage for digital images which were destined to be output and monetised as prints and approached image sharing as a novel loss leader for their real business. In contrast, Flickr joined other photography communities, such as photo.net (1993) and dpchallenge.com (2002), in facilitating and extending the discussion, evaluating and sharing of one's photographic practice with an as yet unknown audience. Flickr's 'groups' feature encouraged discourse and engagement around marginal techniques and cultural forms, from cat scanning to strobography, with contributors fiercely engaged in the conceptual boundaries of the curation of group image pools.[5] Flickr soon became a prominent example of the kinds of communities which might coalesce around distributed media, facilitated by a friendly user interface which made capturing the quotidian moments of one's life both simple and fun. By providing a welcoming home for the 'long tail' of cultural production, Flickr seemed to indicate that even the most banal photographs could attract an audience.

Snapshot culture was a site through which narratives of participation and cultural value could be grasped – both visually and critically. The celebratory narrative of user-generated content was grounded in the vernacular creativity of the Flickr photographer whose work was given new agency through its public visibility. The enthusiasm by which users tagged and uploaded their images underlined the perception that Flickr might evidence the 'wealth of networks' (Benkler 2006) that could harness the cognitive surplus of the user, organising and making meaningful user-generated content. Flickr in turn supported the aims of open culture, through the adoption of creative commons licensing (2004), support for an open API (2004) and the establishment of Flickr commons (2008), which was recently re-launched as the Flickr Foundation (2021). The Flickr community's desire to participate in a creative commons belongs to an optimistic period of the Internet imagined as an open and democratic medium. The souring of that dream in the corporatisation of platforms is directly related to what is being traced here in the flight of the Flickr snapshot from its community to the machine vision dataset.

Financialisation

When Flickr was purchased by Yahoo in 2005, for many, it symbolised the end of a previous era of dot-com decline. And yet, Flickr struggled to find an economic model which would sustain its infrastructure – circulation and scale are interdependent. In the model of venture capitalist funding, photo-sharing platforms like Flickr need to accumulate users to scale, then under the economic conditions of scale, monetise its users through advertising. Yahoo's Marissa Meyer sought to bring in new users to Flickr with the promise of free terabytes of storage, a move which prompted criticism for diluting the existing community. Under Yahoo and Verizon ownership, Flickr launched multiple attempts to monetise, following a logic of platformisation which continues to fetishise 'free' but not 'freedom' for the user. By 2018, Flickr was haemorrhaging money and was 'rescued' by SmugMug, who promised to respect its socio-cultural value, whilst acknowledging the significant economic cost of maintaining its infrastructure. In an interview from the same year, Flickr's new owner Don MacAskill promised:

> We do the lightest amount of data gathering possible. It would be very easy for us to mine data in the photos you've uploaded, generate a profile, target ads on that profile and maybe bundle that data and sell it. I think that's evil. We don't do it and have no plans to ever do it. It's almost web 1.0 advertising. We aren't doing really sophisticated tracking or data mining, or highly targeting ads. We are offering up page view purchases for advertisers to interject into slideshows.
>
> *Shankland 2018*

These assurances about data gathering, albeit in a different circulatory context, were short-lived. Only a year later, Don MacAskill was drawn into controversy when it emerged that IBM had released a facial recognition dataset of 1 million photos uploaded by Flickr users, which was itself a subset of a much larger dataset released by Yahoo Labs in 2014. On NBC News, journalist Olivia Solon (2019) reported what was termed 'the dirty little secret of AI training sets' which had resulted in a situation where 'people's faces are being used without their permission, in order to power technology that could eventually be used to surveil them'. During the same period, researchers Adam Harvey and Jules LaPlace undertook significant investigations leading to the launch of the explosing.ai, a project which traces the aftermath of Flickr's circulation and the incorporation of personal photography into 'image training datasets used for face recognition and related biometric analysis technologies' (Harvey and LaPlace 2021).

The circulation and exploitation of personal photography in AI datasets present a challenge to open culture and offers a compelling example of photo sharing's socio-technical infrastructure, as well as a large gap in public understanding of the photography's role in the networked image economy. Whilst Flickr's part in shaping online photographic culture has been well documented, its position as one of the largest databases of creative commons content might be its most significant, yet largely invisible, contribution in shaping image circulation. What needs looking at in detail is how Flickr become a standing reserve for the computer vision community, in order to get closer to understanding how photographic image is deployed in the networked image culture.

Dataset

On June 24, 2014 – ten years after Flickr was launched, David A. Shamma, a scientist at Yahoo Labs, announced the public release of 'One Hundred Million Creative Commons Images for Research' on his team's blog. The images in question had been uploaded to Flickr between 2004 and 2014 by 578,262 different Flickr users and had been compiled into the YFCC100M dataset: a database representing 99.2 million photos and 0.8 million videos. This was a landmark moment which they anticipated would accelerate research across multiple fields through the provision of 'the largest public multimedia collection to have ever been released' (Thomee et al. 2016, p. 3). The dataset itself did not actually contain any images or videos. Instead, it offered 12 gigabytes of metadata which Flickr had collected as a result of the photo-sharing activities of its users. This included information such as the image's Flickr ID, the identity of its creator, the time it was taken and shared, along with license type, camera type, title, description, user and machine tags, and location data where available. Dynamic metadata (comments, favourites and other social annotations) were not included in the package on the basis they could be retrieved on demand via queries from Flickr's open API. The total size of the multimedia content addressed by the dataset was estimated to be 16.5 terabytes: a staggering amount of data to compute, which its creators suggested would ideally require access to a distributed computing cluster.

The volume of image data that YFCC100M contained eclipsed previous canonical datasets used by computer vision researchers at that time such as MIRFLICKR (2008/2010), which distributed 23,000 and 1 million images, respectively, in its two releases; ImageNet (2009) which offered 14 million photos, PASCAL (2012) 23,000 photos and MS-COCO (2014) 300,000 images. Each of these datasets had been sourced entirely from Flickr, with the exception of ImageNet, of which Flickr URLs account for around one half of the dataset (Malevé 2021, p. 35). In their paper accompanying the release, YFCC100M was described by its authors as a gift to the scientific community in meeting the call for larger, more open and 'diverse' datasets, enabling researchers to build on each other's work through the reproducible and measurable evaluation of algorithms (Thomee et al. 2016). In the seven years since its release, the YFCC100M dataset has become 'a fountainhead from which many derivative datasets were developed' (Harvey and LaPlace 2021).

The rapid gains made by computer vision in the last decade have depended on access to large corpora of images sourced from the Internet. During this period, machine vision researchers established an algorithm's performance is related to volume and regularity of the training data it is exposed to. Models trained on a small subset of photographs sourced in the lab or taken by professionals did poorly when confronted with a 'real-world scene'. In the 1990s, there were substantial limits to the use of photography in image processing scholarship. Access to digital cameras was limited, and as the web was not yet a platform for publishing research, scientists were financially penalised if their journal article or conference paper exceeded the mandated number of printed pages. In the early 2000s, many researchers developed datasets using subsets from a series of 800 stock photography CDs released by Corel, which made benchmarking impossible. Later, photographs were produced in-house or through commissioning professionals. In the 2010s, as image sharing radically expanded, researchers turned to the web as a source of unconstrained imagery possessing a diversity which could not be captured in the lab. In her TED talk on the creation of the canonical dataset ImageNet, Professor Fei-Fei Li describes how the Internet, as 'the biggest treasure trove of pictures that humans have created' offered the variability and scale to train machines to see:

> If you consider a child's eyes as a pair of biological cameras, they take one picture about every two hundred milliseconds, the average time an eye movement is made. So by age three, a child would have hundreds of millions of pictures of the real world. That's a lot of training examples. So instead of focusing on solely better and better algorithms, my insight was to give the algorithms the kind of training data that a child was given by experiences, in both quantity and quality.
>
> *Fei Fei cited in Malevé 2016*

Having conflating camera with eye, photograph with training data, what was required was an abundant source of naturalistic images which could overcome what was perceived as the sampling bias of the professional photographer.[6]

Flickr was an attractive source precisely because of its open API and access to a large store of creative commons licensed data, as well as being a rich source of metadata, from user-generated tags, geo-location information, annotations on 'real-world events', EXIF and values for 'interestingness' which could be mobilised as an aesthetic measure. Crucially, most datasets are sourced from those who do not consider themselves professional photographers. Whilst Flickr is the preferred source for datasets used in object recognition and image retrieval, other photographic communities such as dpchallenge.com and photo.net form the basis of datasets used in aesthetic computing.[7] DPChallenge.com was founded in 2003 with the goal of creating a place where participants 'could teach themselves to be better photographers'. The Photo.net website celebrates its audience as 'photography enthusiasts ranging from newcomers to experienced'.[8] Through this choice of data source, amateur photography gradually became a defining trait of machine vision's photographic culture (Malevé et al. forthcoming).

In computing science literature, the amateur snapshot is therefore positioned as a descriptive and largely transparent image of the world, which in its vast aggregation of lighting, subjects and framing, offers to overcome the limits of professional photographs or those made in the lab. For this reason, images scraped from the web are often characterised as being 'in the wild' – where the Internet is positioned as a source of unknown, uncontrolled, unconstrained imagery operating beyond bounds of human care and attention. This is of particular significance to facial recognition research, where the conventions of professional portrait photography offer up the face as a singular biometric surface, rather than suffering from the noise and occlusions of a body in public space. The dataset creator must align the training data as closely as possible to its real-world application.

From the perspective of YFCC100M's creators, Flickr as a data source provides advantages both in terms of its scale (number of images) and diversity (number of subjects, viewed from multiple angles). In their paper accompanying the dataset, its authors state:

> Our dataset differs in its design from most multimedia collections. The collection of photos, videos, and metadata in **our dataset has been curated to be comprehensive and representative of real-world photography**, expansive and expandable in coverage, free and legal to use, and as such intends to consolidate and supplant many of the existing datasets currently in use.
>
> *Thomee et al. 2016, my emphasis*

But what is this 'real-world photography' that computer scientists curate? In the argument outlined so far, the networked image is relationally constituted in the processual relay between the back-end extractive functions of the database and the front end of user interaction in uploading, sharing and viewing images. By recognising that the snapshot is taken by computer scientists as the measure of seeing and recording the real world, it is possible to understand how a culturally encoded version of the real world is being mobilised as an automated form of objective seeing. This now requires a greater understanding of our knowledge of the snapshot as it crosses between the computer scientists' naturalised view of the snapshot into the discourse of art history.

Amateur Snapshot

From the perspective of computer science researchers, Flickr embodies the kind of photography taken by 'ordinary people', possessing a descriptive banality and unpretentious character which escapes the constructed nature of professional photography or stock photography. Historically, the snapshot has been ascribed to the domain of the amateur − a figure who is apparently unconcerned with issues of composition and technical perfection − and thus produces a fleeting, informal, 'candid' image of the everyday produced through the unskilled use of the automatic camera. Writing in 1996 when the web was in its infancy, Julian Stallabrass described the amateur photographer as lacking 'both an extrinsic social context for their activity and the possibility of financial gain' in contrast to the professional who worked to a defined outcome with expectation of payment, or the category of 'snapper' who 'spends money to make pictures to document holidays, family, friends or special events' (Stallabrass 1996, p. 14). The arrival of social media helped to dissolve the already porous boundaries between those who took snapshots and the committed amateur by aestheticising the everyday and turning users into an engine of vernacular creativity (Rubinstein and Sluis 2008). Here, the position of the amateur was nonetheless celebrated, perhaps, as Stallabrass suggests, because amateur photographers operate as 'artists without pretensions but with a simple faith in their medium... defined by the social and professional uselessness of their work'. In addition, it is a form of photographic production which is uncorrupted by economic imperatives, because the category of amateur photography:

> .. has traditionally been defined against work, especially industrial factory work, where the traces of labour are usually effaced in the process of pro-duction. Taking pictures generally involves travel, wandering, finding things (often by luck), ingenuity and aesthetic appreciation. It is a form of freedom which counters work by commenting on its characteristics of directed, restricted movement and its instrumental relation to objects.
>
> *Stallabrass 1996, p. 26*

The amateur celebrated in Web 2.0 discourse is both a troubling and accommo-dating category for computer vision researchers. Because the quality of training data is central to developing the intelligence of machines, the problem of 'domain expertise' weighs heavily on those tasked with producing datasets at low cost. Image-sharing websites offer the comfort of 'crowdsourced' photography, where its users can be positioned as having an 'investment' in the production and annotation of their images as prolific consumers of photography.[9] In his analysis of the discourse of amateurism in the context of crowdsourcing, Daren C. Brabham (2012) positions the valorisation of the amateur in Web 2.0 as a 'pervasive myth', which is 'not only patently false, but problematic for how we view labour and democratic participa-tion'. In Web 2.0 discourse, the amateur became mobilised as cultural camouflage for the labour of the user, who is alienated from their metadata, whilst providing a model for securing the aggregated byproducts of sociality, which crystallised into a prof-itable commodity. Tellingly, in computer science discourse 'professional' is unprob-lematically characterised as someone who knows how to use their DSLR and has mastery of their controls – it is rarely an economic or creative category.

Whilst there is an established literature crossing art history, cultural studies and memory studies problematising and theorising vernacular photographic snapshot, what is significant to this chapter is the snapshot's relation to automatism and its capacity to render the photograph transparent and position the everyday as calcul-able. The snapshot has been historically positioned as an authentic and privileged take on everyday life, eschewing the constructed and commercially tainted realm of professional photography. Unmonumental, fleeting and semi-automated, the snap-shot is a product of the humble tools available to the amateur photographer who does not attempt to manipulate the scene or the subject. In Aperture's 1976 issue on the snapshot, the artist Lisette Model embodies this position, stating:

> I am a passionate lover of the snapshot, because of all photographic images it comes closest to truth. The snapshot is a specific spiritual moment. It cannot

be willed or desired to be achieved… A snapshot is not a performance. It has no pretence or ambition… Innocence is the quintessence of the snapshot. The professional photographer, in spite of the instantaneous and spontaneous means at his disposal, can never achieve that degree of innocence.

Green 1974, p. 4

Julia Hirsch (cited in Chalfen 1983) further connects the impulsive automatism of the snapshot to the realm of psychoanalysis, stating 'Candid photographs take us into the realm of abstraction, of subconscious … like inkblots or obscure poems'. From this perspective, the naïve camera operator produces a spontaneous rendering of an unmediated real, operating outside cognitive bias through the unschooled use of the automatic camera.

Camera-phone photography serves to reinforce and extend the automatism of the snapshot, whilst the logic of social media creates the conditions of a repetitive and habitual photographic practice. The arrival of the photoblogger in the early 2000s pushed the logic of passive snapshot production further, in which the wearable automatic camera might further collapse the boundaries between photography and the world. It is perhaps not surprising to learn the YFCC100M dataset contains over 200,000 contributions from Andy E. Nystrom of Canada, whose Flickr feed of over 6 million images presents the output of his Canon Powershot SX60 at a rate of one image per second (Figure 2.1) as he navigates Shelbourne Plaza, or wanders

FIGURE 2.1 Flickr feed of Andy E. Nystrom, IMG_8534 – IMG_8519, November 2021. All Images Courtesy Andy E. Nystrom CC BY-NC-ND 2.0

through Uptown Shopping Plaza, or as he watches movies at home amongst empty beer bottles and an adoring cat.[10] A Seattle trip in 2009 generates 1558 photos but only 38 views on Flickr yet secures him the privilege of being one of the dataset's 578,262 photographers to receive a citation in the computing science paper authored by YFCC100M's creators (Thomee et al. 2016).

It is, however, important to remember that, as Chalfen (1983) points out, the snapshot is far from neutral: it was camera manufacturers who promoted the 'snapshot view of the world' together with other 'conventional views produced by professional studio photographers, photojournalists and fine art photographers'. Today, the snapshot aesthetic is mobilised as 'an intentional style' by brands to communicate 'honesty, sincerity, and spontaneity'. Jonathan E. Schroeder (2013) explains:

> .. realistic looking snapshots often seem to have no style at all, or appear unstylized. This "absence of style" is itself a style, and the contradiction between ads that don't appear stylized and the creative stylization required to produce them remains at the heart of snapshot aesthetics' strategic success.

In cannibalising the naïve realism of the unstylised snapshot, the dataset can, as a cache of networked images, be 'camouflaged as a non-political, non-significant and non-ideological site that does not merit textual analysis' (Rubinstein and Sluis 2008, p. 23). The wild scales of the Flickr snapshot promises to cure the dataset of its representative (but not representational) bias, operating as the 'other' to the professionally produced image: without the endless repetition of tropes, captured from multiple angles, times of day, variations and occlusions, the ability of machines to extract patterns is diminished and prejudice creeps in.

But perhaps most significantly, the naïve snapshot offers the potential to bypass the cognisance of the camera operator, which might prejudice a truthful representation. The automatism of the camera, which produces the spontaneous snapshot, is positioned as that which renders a transparent image of the world in a millisecond exposure. This is consistent with a model of cognition mobilised elsewhere in the machine vision pipeline, where the economic pressures of labelling datasets by annotators employed on crowdsourcing platforms privilege the millisecond glance. For computer vision researchers, this means the snapshot can enter the lab as a visual stimulus which produces new taxonomies (see Malevé, this volume) or enters the dataset as an unmediated product of a visual stimulus, as I argue here.

Real-World Photography

Crucially, the 'real' that is mobilised in the 'real-world photography' of the dataset is functionally very different from the real which has haunted photography theory. Computer science papers remain untroubled by the relation between the photograph and its referent, or the digital malleability of the jpeg nor its polysemy. The 'real' of the dataset is statistical and calculable, in which real-world photography is ontologically aligned to real-world variations, and representative of the multiplicity and complexity of real-world scenes. The extraction of millions of photographs from the automated shutters of millions of lenses becomes the method through which dataset curators might capture the long tail of the 'real'. It is telling then, that in verifying the YFCC100M dataset's representativeness as real-world photography, its creators measured it against Flickr itself. Having downloaded a random sample of 100 million public Flickr photos, they compared the 'relative frequency with which content and metadata were present in each' and concluded their dataset was representative, albeit with a slight bias towards DSLR cameras in YFCC100M (Thomee et al. 2016, p. 5).

In this respect, a collection of 10,000 photos of sunsets, cats, clocks or clouds therefore possesses both statistical variance and iconic unity which is required by the dataset. The snapshot is representative precisely because, as a cultural form, it 'embodies the tension between individuality and generality, between originality and conformity' (Berger 2011, p. 182). In her defence of the amateur snapshot, art historian Annebella Pollen (2018) attempts to liberate the snapshot from the category of mindless cliché, and in particular, the mass production and circulation of sunsets. In emphasising the individual character or specificity of each sunset photo – she inadvertently illuminates mass photography's significance to the dataset, arguing:

> The continuing production and circulation of sunsets may seem… to suggest a brainless, sheep-like adherence to popular image templates, and yet, for all photography's reproducibility, each photograph is always unique to the photographer; they were each individually moved to record it… Patterns exist because they show what matters to people… Sunset photographs may all look the same, but the meaning changes with each one.
>
> *Pollen 2018, p. 87*

This point is echoed by art historian Geoffrey Batchen (2008, p. 125) who points out that when examining vernacular photography 'you'll discover that each example captures a unique pose, even if that pose obediently repeats a million other, very similar poses. They are all the same, but they are all also just slightly different from each other'. Here we see also the logical endgame of Villem Flusser's (2000, p. 26) contention that the role of the photographer is to 'exhaust the photographic program' of the camera apparatus by realising the 'possibilities not yet discovered within it' in the pursuit of 'new possibilities of producing new information'. This becomes crucial for machine learners that must comprehend or 'image' the category of sunset through a combination of iconicity and aggregated repetition. For this reason, the average amount of photographs per category in ImageNet is over 10,500 (Fei-Fei cited in Malevé 2021, p. 35), whilst YFCC100M offers 639,453 photographs of sunsets, of which 189,440 are 'amazing sunsets', 189,44 are 'beautiful sunsets' and 140,342 are 'awesome'.

The shift from the singular to the collective viewpoint, embodied by aggregation of images, produces a 'framelessness' which Mark Andrejevic (2020) argues is characteristic of big data. As the creators of YFCC100M argue:

> Photos and videos provide a wealth of information about the universe, covering entertainment, travel, personal records, and various other aspects of life in general as it was when they were taken. **Considered collectively, they represent knowledge that goes beyond what is captured in any individual snapshot and provide information on trends, evidence of phenomena or events, social context, and societal dynamics**.
>
> *Thomee et al. 2016, 1 my emphasis*

Once dissolved into the logic of the dataset, extracted from the specificity and sociality of the interface, the snapshot is given a 'framelessness' and scale which appears natural. As Andrejevic explains, framelessness is set in contrast to representations because 'representations are always necessarily selective, biased, and therefore subject to debate, correction, and disbelief' (2020, p. 263). In this respect, Andrejevic notes, 'the convention of objectivity works to dissimulate the existence of a frame by implying that the selection of facts and their presentation in a story takes place neutrally: that the world is being presented simply "as it is"' (2020, p. 261). This is reflected in the concern for 'representativeness' over 'representation' in securing the scientific legitimacy of the dataset.

In contemporary culture we now reach a situation where the amateur snapshot is computationally valorised for its statistical variation, whilst culturally derided for its embrace of repetitive banalities. The amateur photographer is not only the producer of clichés but also a gateway to an imaginary 'real' which can be grasped through scale and variation. The snapshot possesses the unconstrained automatism required of the dataset, in which the quotidian, habitual, ambient production of images by naïve photographers has the capacity to capture the long tail of cultural production and, by extension, the real. The bias of the photograph, which the machine vision researcher must contend with, is limited to its statistical variability, in which the wild scale of Flickr's 'real-world photography' is set against the staid conventions of the professional photographic studio. Bias is not understood in terms of the politics of photographic representation nor the subjectivity of the photographer, both of which have been rendered transparent. Instead it is a calculable property in which the framing of a single cat photograph might be obliterated or absorbed into the generality produced by the semi-automated or compulsive output of millions of cat photographers.

Here the cultural (photograph) and computational (statistical) collide: the amateur photograph is culturally conservative, but for machine learning, the professional photographer is too conservative. Crucially, the amateur snapshot offers a cultural form which enables computer scientists to translate the everyday into the statistical, whilst the amateur snapshot's status as hidden, devalued or of low cultural value lends itself to processes of extraction. From this position, we reach a new socio-technical view of objectivity: in which the naïve snapshot in all its grainy, low quality is positioned as an authentic fragment of the real, whilst its ubiquity brings with it the perceived objectivity of capturing diversity, the long tail of sociality, and the generality of photography.

This is a point made forcibly by Sarah Kember (2018) who suggests the concerns of big tech are less about empowering users but 'staking direct claims to sociality and to everyday environments constituted by users and intelligent artefacts alike'. The networked image, a product of distracted engagement with the everyday operates as an invisible, mundane and slightly silly image used as visual stock for AI industries because, as Kember (2018, p. 226) argues:

> The quotidian, the vernacular is thus contested ground and amateurism is a route to an automated, ambient, animated and augmented existence designed according to the imperatives not of technology per se, but of surveillance-based markets.

The derided qualities of the amateur snapshot: its banality, ephemerality, insignificance, clichés are precisely the values that are valorised and operationalised in the machine learning pipeline.

Calculable Surface

From here, it becomes clear how Flickr operates in three different registers: firstly, as a site of photo sharing and sociality, in which communities support the circulation of their work through open licensing. Secondly, it is searchable database of image content, supported by rich metadata, dissociated from its origins with an accessible API facilitating third-party access. Thirdly, it can be considered an 'epistemic harvest' of metadata in the form of the dataset. This epistemic harvest is what Francis Hunger (2018, p. 56) calls 'a new regime of data production: a documented interpellation and recorded extraction from every participant in the social field' in which 'each action, even the seemingly non-productive action… has turned into an act of data production'. Photography promises to materialise the social world as data, in which the iconicity of the image operates as a lure (Hoelzl and Marie 2015) which can be mobilised for further metadata generation. This is reflected in the language of computing science papers, which describes how to deploy 'large-scale image mining' and 'leverage Flickr groups' in order to build 'comprehensive visual resources', which can be 'exploited for image retrieval' (Ginsca et al. 2015).

In our 2008 article, Daniel Rubinstein and I drew upon the work of Paul Frosch, whose work on stock photography industries offered an analysis of the 'rhetorics of the overlooked' in which commercial images are engineered to be unremarkable and non-ideological, in which 'indexical singularity' is erased in favour of 'uniformity and recurrence' (p. 189). Thirteen years later, this convergence of snapshot and stock image is made explicit in not just the dataset but also feeds back into the business models of contemporary photo-sharing platforms. One prominent example is the tech start-up and photo-sharing platform EyeEm, which launched in 2011 as an alternative to Instagram. During its first years of user expansion, it promoted itself as a platform that genuinely cared for the photographic community and embarked on an ambitious cultural programme, throwing parties, workshops, conferences, exhibitions and competitions in collaboration with leading cultural and technology organisations. Under the inevitable pressure to monetise, it pivoted to a stock photography company, allowing its community to earn money from their image-sharing labours, whilst publicly promoting itself as a gateway 'authentic' stock imagery not found on other platforms, in yet another gesture towards the 'innocence' of non-professional imagery. With its purchase of machine vision start-up sight.io in 2014, EyeEm pivoted from stock agency to AI company. Today, EyeEm leverages the photographs shared by its community simultaneously as stock photography and also as a dataset to train machine learning algorithms in aesthetic evaluation. The AI model is then sold on as a service to brands looking for image stock which matches the aesthetic values of their brand.

As the example of EyeEm and Flickr shows, the radical decontextualisation of computational systems has turned all personal photography into the 'found photograph' – extracting it from the network of relations, interactions and contexts, which binds it to the social world. In the gap between Flickr as photo-sharing community and Flickr as dataset, it becomes possible the glimpse the incommensurability of image as representation and image as data. The contemporary harvesting of Flickr data is at odds with the ideals of a sharing, open and collaborative epistemic community, in which the extractivism of contemporary AI industries now threatens the mobilisation of the web as knowledge commons. This is a point made by Matteo Pasquinelli and Vladan Joler (2020, p. 5) who observe that Lawrence Lessig, from the position of the early 2000s, 'could not predict that the large repository of online images credited by Creative Commons licenses would a decade later become an unregulated resource for face recognition surveillance technologies'. The open licenses under which Flickr snapshots were given consent to be used cannot be revoked, and the research of Adam Harvey and Jule LaPlace (2021) have shown how such images can be unwillingly incorporated in biometric systems which might be used against these same communities. They point out that even if biometric datasets sourced from YFCC100M are taken down from a corporate server, they remain stored across many millions of researchers' machines or continued to be shared via academic torrents.

From the perspective of the dataset, the networked image aligns with what Mark Andrejevic calls 'the fantasy of total information collection' which informs twenty-first century automated media in 'promising to bridge the gap between representation and reality' (2020, p. 81). The networked image vacillates between fluidity and categorisation, between polysemy and classification, between the amorphousness of the big data cloud and the specificity of someone's actual cat. The networked image demands curation, stabilisation, semiotic stability and taxonomic clarity. The paradox of the networked image is that whilst it appears to promote slippage and celebrate movement, in which users participate in the life of the image, it produces a ubiquity which must be contained by its foreclosure into massive databases managed by the tools of big tech at massive economic and environmental cost. The cultural stabilisation of the networked image as 'amateur' or 'professional' or 'snapshot' has epistemological significance. It enables an enforcement of aesthetic boundaries and socio-economic values onto the mass of image data. The photographic pipeline of machine learning seeks to stabilise and align the inherent polysemy of the image in order to produce actionable insights. In doing so, it renders the photograph transparent.

Notes

1 For a discussion of this condition of circularity with respect to dataset curation see Malevé et al., forthcoming.
2 For a comprehensive overview of these debates, see Birhane and Pabhu (2021).
3 Canonical videos in this genre include Micael Widell's 2021 video *Instagram Betrays Photographers, Flickr Is Dead, 500px Sucks. Where to Go Now?* www.youtube.com/watch?v= b5FHv0g0agg and Ted Forbes' 2020 video *Flickr Wants Your $* www.youtube.com/ watch?v=vtaN-ZXfrVg with around 145k views and 1300 comments between them.
4 For a discussion of AI and curating in the context of the machine learning pipeline, see Malevé et al. (forthcoming).
5 For a discussion of the Somebody Else's Cat Flickr group, see Lop Lop (2012).
6 For Fei-Fei Li, the sampling bias can be averaged through images in the wild. As she explains: 'The Google image search engine largely return[s] images found on people's personal websites, most often taken with a snapshot camera. Although everyone has a bias when taking a picture, we believe that the large number of images from different unknown sources would help average out these biases.' Fei-Fei et al. (2007).
7 For a discussion of aesthetic datasets and their curation, see Part 4 of Sluis (2019).
8 Datasets commonly used in aesthetic computing have been harvested from photography communities including behance, photo.net, gurushots and dpchallenge.com.
9 This is particularly true of aesthetic datasets such as AVA, which I discuss further in Sluis (2019).
10 Andy Nystrom's Flickr feed of over 6 million images can be viewed at www.flickr.com/ photos/24917258@N05/.

Bibliography

Allspaw, John and Paul Hammond. 2009. "Velocity 09:'10+ Deploys Per Day: Dev and Ops Cooperation at Flickr.'" www.youtube.com/watch?v=LdOe18KhtT4.
Andrejevic, Mark. 2020. *Automated Media*. London; New York: Routledge.
Batchen, Geoffrey. 2008. "Snapshots." *Photographies* 1 (2): 121–42. https://doi.org/10.1080/ 17540760802284398.
Benkler, Yochai. 2006. *The Wealth of Networks: How Social Production Transforms Markets and Freedom*. New Haven, CT: Yale University Press.
Berger, Lynn. 2011. "Snapshots, or: Visual Culture's Clichés." *Photographies* 4 (2): 175–90. https://doi.org/10.1080/17540763.2011.593922.
Birhane, Abeba, and Vinay Uday Prabhu. 2021. "Large Image Datasets: A Pyrrhic Win for Computer Vision?" In *Proceedings of the IEEE/CVF Winter Conference on Applications of Computer Vision*, 1537–1547.

Brabham, Daren C. 2012. "The Myth of Amateur Crowds: A Critical Discourse Analysis of Crowdsourcing Coverage." *Information, Communication & Society* 15 (3): 394–410. https://doi.org/10.1080/1369118X.2011.641991.

Carrington, David. 2020. "How Many Photos Will Be Taken in 2020?" Mylio Blog. 29 April 2021. https://blog.mylio.com/how-many-photos-will-be-taken-in-2020/.

Chalfen, Richard. 1983. "Exploiting the Vernacular: Studies of Snapshot Photography." *Studies in Visual Communication* 9 (3): 70–84.

Flusser, Vilém. 2000. *Towards a Philosophy of Photography*. London: Reaktion Books.

Ginsca, Alexandru Lucian, Adrian Popescu, Hervé Le Borgne, Nicolas Ballas, Phong Vo, and Ioannis Kanellos. 2015. "Large-Scale Image Mining with Flickr Groups." In *MultiMedia Modeling*, edited by Xiangjian He, Suhuai Luo, Dacheng Tao, Changsheng Xu, Jie Yang, and Muhammad Abul Hasan, 8935:318–34. Lecture Notes in Computer Science. Cham: Springer International Publishing. https://doi.org/10.1007/978-3-319-14445-0_28.

Green, Jonathan. 1974. "Aperture: The Snapshot (1974)." *Aperture* 19 (1): 1–128.

Harvey, Adam and Jules LaPlace. 2021. "Exposing.Ai." https://exposing.ai.

Hoelzl, Ingrid and Rémi Marie. 2015. *Softimage: Towards a New Theory of the Digital Image*. Bristol: Intellect.

Hunger, Francis. 2018. "Epistemic Harvest." *A Peer-Reviewed Journal about Research Values* 7 (1): 167.

Kember, Sarah. 2018. "The Becoming-Photographer in Technoculture." In *Photography Reframed: New Visions in Contemporary Photographic Culture*, edited by Ben Burbridge and Annebella Pollen. London: I. B. Tauris & Company, Limited, pp 213–220.

Lop Lop, Dr. 2012. "Somebody Else's Cat: A Study in the Protohistory of the Internet Cat Meme." *The Photographers' Gallery: Unthinking Photography*. https://unthinking.photography/articles/somebody-else-s-cat-a-study-in-the-protohistory-of-the-internet-cat-meme.

Mackenzie, Adrian and Anna Munster. 2019. "Platform Seeing: Image Ensembles and Their Invisualities." *Theory, Culture & Society* 36 (5): 3–22. https://doi.org/10.1177/0263276419847508.

Malevé, Nicolas. 2016. "'The Cat Sits on the Be,' Pedagogies of Vision in Human and Machine Learning." Unthinking Photography: The Photographers' Gallery. https://unthinking.photography/articles/the-cat-sits-on-the-bed-pedagogies-of-vision-in-human-and-machine-learning. Accessed 6 December 2021.

Malevé, Nicolas. 2021. *Algorithms of Vision. Human and Machine Learning in Computational Visual Culture*. PhD Thesis. London South Bank University.

Malevé, Nicolas, Katrina Sluis, and Gaia Tedone forthcoming. "Curating in the Wild" In *Curating Superintelligences: Speculations on the Future of Curating, AI and Hybrid Realities*, edited by Joasia Krysa and Magdalena Tyżlik-Carver. London: Open Humanities Press.

Miceli, Milagros, Martin Schuessler, and Tianling Yang. 2020. 'Between Subjectivity and Imposition: Power Dynamics in Data Annotation for Computer Vision'. *Proceedings of the ACM on Human-Computer Interaction* 4 (CSCW2): 1–25. https://doi.org/10.1145/3415186.

Pasquinelli, Matteo and Vladan Joler. 2020. "The Nooscope Manifested: AI as Instrument of Knowledge Extractivism." *AI & Society*, November. https://doi.org/10.1007/s00146-020-01097-6.

Pollen, Annebella. 2018. "When Is a Cliché Not a Cliché: Reconsidering Mass-Produced Sunsets." In *Photography Reframed: New Visions in Contemporary Photographic Culture*, edited by Ben Burbridge and Annebella Pollen. London: I. B. Tauris & Company, Limited, pp 82–88.

Rubinstein, Daniel and Katrina Sluis. 2008. "A Life More Photographic: Mapping the Networked Image." *Photographies* 1 (1): 9–28. https://doi.org/10.1080/1754076070 1785842.

Rubinstein, Daniel and Katrina Sluis. 2013a. "Notes on the Margins of Metadata: Concerning the Undecidability of the Digital Image." *Photographies* 6 (1): 151–8. https://doi.org/10.1080/17540763.2013.788848.

Rubinstein, Daniel and Katrina Sluis. 2013b. "The Digital Image in Photographic Culture: Algorithmic Photography and the Crisis of Representation." In *The Photographic Image in Digital Culture*, edited by Martin Lister, 2nd ed., 22–40. London: Routledge.

Schroeder, Jonathan E. 2013. *Snapshot Aesthetics and the Strategic Imagination*. Rochester, NY: Social Science Research Network. SSRN Scholarly Paper ID 2377848. https://papers.ssrn.com/abstract=2377848.

Shamma, David A. 2014. "One Hundred Million Creative Commons Flickr Images for Research." Tumblr. Yahoo Research (blog). https://yahooresearch.tumblr.com/post/89783581601/one-hundred-million-creative-commons-flickr-images-for.

Shankland, Stephen. 2018. "Despite the Blowback, Flickr's CEO Happy with Limits on Free Photo Sharing." CNET. 10 November 2018. www.cnet.com/tech/services-and-software/flickrs-new-limit-on-free-photo-sharing-is-helpful-not-hurtful-ceo-says-q-a/.

Sluis, Katrina. 2019. "Photography Must Be Curated!" Still Searching: Fotomuseum Winterthur (blog). 15 September 2019. www.fotomuseum.ch/en/explore/still-searching/series/156409_photography_must_be_curated.

Solon, Olivia. 2019. "Facial Recognition's 'Dirty Little Secret': Social Media Photos Used without Consent." *NBC News*. www.nbcnews.com/tech/internet/facial-recognition-s-dirty-little-secret-millions-online-photos-scraped-n981921. Accessed 13 December 2021.

Stallabrass, Julian. 1996. "Sixty Billion Sunsets." In *Gargantua: Manufactured Mass Culture*. London: Verso, pp 13–39.

Thomee, Bart, David A. Shamma, Gerald Friedland, Benjamin Elizalde, Karl Ni, Douglas Poland, Damian Borth, and Li-Jia Li. 2016. "YFCC100M: The New Data in Multimedia Research."

3

POST-CAPITALIST PHOTOGRAPHY

Ben Burbridge

If the egalitarian possibilities of the image emerge most fully at the moment when what was once called photography collides with networked computation, to be reconfigured as what is sometimes called the 'networked image', they do so at a time when the material realisation of this potential faces unprecedented challenges. Where the social production, circulation and consumption of the networked image are marked by principles of sharing, abundance and connection, the corporations that monetise interactions through the ownership of platforms and harvesting of data seek instead to mine, extract and hoard. The phoney utopianism peddled by Zuckerberg, Getty and Dorsey deploys one as ideological cover for the other: access is fetishised, economics disavowed. By approaching the relationship through the lens of post-capitalism and the politics of work, this chapter asks instead if the networked image's social condition might be extended to the economic realm.

Post-capitalist photography is a method, an idea and a speculative future. It names a series of possibilities linked to a radically democratic apparatus: the sharing of experiences that photography makes possible, and the experience of making, viewing and publishing photographs as something we share. It is not new, existing across the radical histories of photography, where writers and artists have linked access, mass production and replication to democracy, egalitarianism and socialism (Benjamin 2008; hooks 2003; Moholy-Nagy 1987; Spence 1995). But neither is it devoid of novelty, given its relationship to networked digital cultures where old hierarchies collapse and photography as a mass practice gains new forms of visibility. To understand its potential, along with the cultures it frames and is framed by, it is necessary to think about the networked image laterally and historically, particularly in terms of photography's monetization and the labour on which it relies.

This chapter addresses that task via four core concerns. First, I consider changes to value production that accompanied photography's integration into computational systems. It is by now well known that the primary economic functions of

DOI: 10.4324/9781003095019-5

the image are linked to the algorithmic surveillance of online interactions. 'Post-capitalist photography' is thus a specific manifestation of a more general shift in perspective suggested by the term, the 'networked image', linking 'networked' to a radically expanded conception of photography. The representational content of the image – for so long the primary focus of scholarship about photography – is indivisibly wedded to the technical, social and political infrastructure that sustains it. Grasping this point is the first step towards a strategic navigation of the network, cultivating the forms of critical literacy that the image's networked condition facilitates and demands. When economic value is determined by the algorithmic surveillance of our online lives, those lives continue to be shaped in fundamental ways by the visual content of images.

Secondly, I explore how relationships between subjectivity and labour play out in relation to networked image cultures: in terms of the 'mandatory individualism' and performative selfhood of post-Fordism (Fisher 2016); but also the more radical forms of identification that have, and could yet, grown out of our socially mediated interactions. Drawing on Jodi Dean's (2016) idea of 'selfie communism', I consider the paradoxical ways in which the realisation of individual selves via the sharing of images may help to loosen photography's historical attachment to private property, particularly in the context of its domestic uses. The 'Kodak moment' once signalled the preservation of experience for consumption within the closed confines of the nuclear family. Today our snapshots are freely shared with audiences of we don't know how many. It is in this sense that the networked image potentially extends the logic celebrated by Walter Benjamin's famous essay about mechanical reproduction. When images exist simultaneously on an indeterminate number of screens, could this help to fundamentally reconfigure our relationship to property, to work and to each other?

Thirdly, the chapter considers the different strategies via which such principles could be translated into material change. Here it draws insight and inspiration from various discussions of a world after capitalism that have entertained the imaginations of writers and activists in recent years. The term 'post-capitalism' has been enlisted to describe two related projects. Both have implications for our understanding of the networked image. The approach associated with feminist economists J.K. Gibson-Graham focuses on real-world alternatives to neoliberalism, particularly cooperative ventures and community economies (Gibson-Graham, 1996, 2006). Writers including Nick Srnicek and Alex Williams (2016), Mark Fisher (2019) and Paul Mason (2016) have instead set out to envisage larger scale structural changes, linking collectivist politics to a post-work world. With the provision of a universal basic income (UBI), they argue, society could be guaranteed a level of material security that would enable people to pursue their creative ambitions while avoiding the drudgery of waged labour.

In the context of my argument, the introduction of UBI – particularly a UBI funded via taxes levied against platform's corporate owners – would provide an important mechanism via which the potential of post-capitalist photography could be pursued: extending principles of play, abundance and sharing to the economic domain. This identification of political horizons serves two important roles. Approached as what Helen Hester (2017) calls an 'affective technology', thinking about alternative futures provides the space to hope and – by insisting that what happens next need not be the same as what happened before – underlines the plasticity of the current order. The identification of horizons also enables objectives to be clarified and directions of political travel to be defined.

My conclusion insists on post-capitalist photography as a necessarily global project, linking the work of social media 'prosumers' to the labour involved in the production of hardware and support work, whether smartphone manufacture in Shenzhen or the shady world of content moderation. My goal here is not so much to assemble a list or itinerary, but to plot a constellation, exploiting the ubiquity of networked images to 'cognitively map' a global economic system and, in so doing, render it legible at the level of individual subjects (Jameson 1987). The capacity to situate manual and cognitive labour in relation to each other provides an initial step towards identifying the need for truly global forms of solidarity, pitched against the extractivism that too often powers the production, circulation and consumption of the networked image.

A word, finally, about my terminology. My continued use of 'photography' – always, if sometimes by implication, in the context of 'post-capitalist photography' – is rooted in three considerations. First, the practices on which I focus are linked to cultural phenomena that I suspect the majority of people – particularly those outside academia – continue to describe as photographs and/or photography. My goal is to intervene in an already available space and make use of an existing container, in ways that both expand and reframe its current contents. The familiarity and ubiquity of the term, like the familiarity and ubiquity of what it typically describes, possesses profound political utility in ways that I suspect are not true, or not true in the same way, as a term such as 'the networked image'.

Secondly, my objectives are primarily, if not exclusively, political and economic. The goal is not so much to understand socio-technological shifts to image culture, or to name the ontological changes they accompany, but instead to think about the political and imaginative possibilities they bring into view. Accepting the widely used term, despite its many imperfections, appears the most effective way to harness the potential of photography, a potential signalled by the prefix 'post-capitalist' that accompanies it in the context of this chapter.

Thirdly, the political possibilities that post-capitalist photography seeks to name are anchored in what Robin Kelsey and Blake Stimson (2008) have described as the 'promise of photography'. Across its histories, photography has mediated between internal and external domains, between the realm of subject and object. Harnessed as a detached observer by science, as persuasive witness by journalism or as confessional subject by art, photography enabled specific acts of viewing to be experienced publicly and collectively. The framing of a pocket of time and space by makers situated in a specific pocket of time and space was reproduced and disseminated across other times and spaces.

This is at once an ontological and philosophical assertion. What is often called 'photography' today relies on a computer simulation of the 'iconic indexicality' that, for many, has most defined photography historically. Yet the sheer volume of photographs being taken and shared, and the immediacy with which they can be viewed – two changes typically conjured by talk of photography's current 'ubiquity' – have arguably reproduced and reinforced its relation to the real, even as the ontological condition of the digital image unsettles that characterisation (Hand 2012). When someone takes and then immediately views a selfie on a smartphone, it is experienced not as a computer simulation of photography, but as a photographic representation of the pocket of time and space in front of the camera. When the image takes flight across the network – a default position that today is always already assumed – two forms of sharing take shape in relation to each other: reality as image, and the image in reality. This chapter contends that photography's promise could yet be reconfigured and revived by what Stimson and Kelsey describe as its 'sociality', even if – in doing so – photography is reshaped and, indeed, rechristened as the 'networked image'. Post-capitalist photography names a specific political potential at the core of that observation and suggests it exemplifies the larger principles that animate utopian political thought. It is through photography's interactions with the network that the need to remodel its social relations in the image of a shared culture comes into focus with particular clarity.

Attention Economies

The integration of snapshot photography into networked computational systems has helped shift the primary site of its monetisation. This is best understood as a chain or process, beginning with the purchase of the camera. In pre-digital times, people would buy the camera and pay both for the film required to take photographs and to have the film processed. The profitability of mass photography lay primarily in the production, sale and developing of film, which – unlike cameras – were continual, not one-off, purchases. Thus, Kodak adopted what is sometimes referred to as a 'Gillette' model. Its cameras were priced cheaply, much like Gillette priced its razors. Money was made through film, in the same way, Gillette profits from razorblade sales.

People continue to pay for cameras today, many of them included in smartphones. But we no longer purchase film nor pay to have that film developed. New sites for monetisation have emerged to offset that loss, the majority based on the public sharing of photographs across computer networks and the online interactions they facilitate. We often pay monthly tariffs to access the network, sometimes impacting on the production of value earlier on in the chain, with the cost of smartphones discounted in order to attract us onto particularly monthly deals, for example. There are parallels to be drawn with a Gillette model here, but it is network access, not film and developing services, that we pay for. The two are not directly comparable; however, where Kodak was a 'vertically integrated' company and had financial stakes in almost every stage of the photographic process, smartphones are generally produced by companies other than those which provide broadband. The main exception today is Apple, which builds hardware, owns the core software experience, optimises its software for that hardware, equips it with its own web services and controls the selling experience through its own retail stores.

The primary economic value of the networked image is linked to opportunities for targeted advertising. Advertisers pay to access the attention generated by platforms where photographs circulate, and to target advertising based on insights made available through the algorithmic surveillance of online interactions. As Mark Andrejevic explains:

> The goal is not just to determine what information might be useful to consumers, but how best to trigger the anxieties and concerns that might motivate them to buy, how best to use information about their hopes, dreams, and desires, their moods and their health, as well as their romantic and family histories, to figure out how to bend consumer behaviour to the priorities of marketers…all of our activities, to the extent that they can be redoubled in the form of data harvested by interactive networks, return to us in unrecognizable, perhaps even unremarked form.
>
> *Andrejevic 2011*

'Free' services such as non-commercial use of Getty's vast stock image library are powered by similar principles. The data generated about user's behaviour – and particularly the resulting capacity to predict 'future image trends' – is far more valuable than royalties paid for the use of individual images (Sluis 2014).

It has become increasingly common to think about networked photography in relation to what artist Harun Farocki (2004) called the 'operational image'. Interactions with photographs are viewable to machines as data, with behaviour subject to constant corporate scrutiny. Images not only 'represent an object' but are also 'part of an operation' (Hoelzl and Marie 2014, p. 247). The likes of Instagram, Google Street View and Getty Images push that logic to new extremes. It is precisely *because* we experience the photographs as images that we so often ignore the larger operations in which photography is enmeshed. For the photographic historian Geoffrey Batchen,

> …the invisibility of the photograph, its transparency to its referent, has long been one of its most cherished features. All of us tend to look at photographs as if we are simply gazing through a two-dimensional window onto some outside world. This is almost a perceptual necessity; in order to see what the photograph is of, we must first repress our consciousness of what the photograph is.
>
> *Batchen 2000, p. 263*

Batchen highlights those ways in which the representational content of photographic images makes particular appeals to our attention, and how this can make it difficult to recognise what the photograph is in material terms. This can also make it difficult to develop a strategic and critical understanding of what Katrina Sluis (2018) describes as 'the techno-social infrastructures which sustain the photographic image'. It is in this sense that photography has long exemplified the operative logic of communicative capitalism today, insofar as its users have typically focused on representational content visible on the surface of the photograph – at its visual interface – not the materiality of the photographic object or the systems of production and circulation in which photography is entwined.

Why does this matter? In his book *24/7*, Jonathan Crary (2013, pp.47–8) describes the effects of the network as the realisation of earlier industrial fantasies of round-the-clock productivity. The capacity of networked image culture to produce capital around the clock links it to a world 'reimagined as a non-stop work site or an always open shopping mall of infinite choices, tasks, selections, and digressions'. What is described here is something more than a post-Fordist erosion of distinctions between the professional and the private. There are fewer and fewer times when we are not producing in the interests of capital. Crary identifies what Foucault called a 'network of permanent observation', but one that sees most of the 'historically accumulated understandings of the term "observer"... destabilized... that is, when individual acts of vision are unendingly solicited for conversion into information that will both enhance technologies of control *and* be a form of surplus value'.

Trebor Scholz (2012) has suggested that people sometimes find it hard to understand social media activity as unwaged work 'because in opposition to traditional labour, casual digital labour looks merely like the expenditure of cognitive surplus, the act of being a speaker within a communicative system'. Nevertheless, our 'intimate forms of human sociability are being rendered profitable'. Oligarchs 'capture and financialise our productive expression and take flight with our data. We, the "users" are sold as the products' (Scholz 2012, p. 2). For Nick Srnicek, data is better understood as 'the raw material that must be extracted, and the activities of users to be the natural source of that raw material'. Like oil, 'data are material to be extracted, refined and used' (Srnicek 2017, pp. 39–41).

It is difficult to imagine a situation in which the users of social media platforms could unionise to demand better working conditions, or to have a system for payment introduced; much simpler to imagine opting out of social media altogether, or developing alternative platforms that do not rely on data mining and advertising for revenue, in ways closer to consumer activism than to industrial action. But when data is understood as a resource in need of refining, it does not appear to be 'natural' in the same way as crude oil because it involves the monetisation of human interactions. It is the social dimension of the process that lends it the appearance of unpaid work, even if the nature of value production means it does not sit comfortably within traditional definitions of labour (Beech 2019).

It is not only data mining that should concern us here but also the distribution of the resulting wealth. The financial arrangements of Instagram's owners, Facebook Inc., are typical of many other multi-nationals, which exploit de-territorialised activities to avoid taxation. For many years, Facebook based its international operations in Ireland, where corporate tax rates were set at 12.5 per cent. This compared favourably to Britain, where businesses at the time were taxed at 19 per cent (still low compared to other countries). Until 2017, money made from sales to advertisers in the United Kingdom was routed via Ireland. Thus, an additional 6.5 per cent of Facebook's annual UK revenue went into the pockets of shareholders, rather than being contributed towards the healthcare, education and other public services on which many Facebook users rely. The sum was actually much larger due to the corporation's use of staff bonuses and the payment of hefty licensing fees to subsidiaries registered in the Cayman Islands, which reduced the level of profits on which it was required to pay tax in Ireland.

Critical Literacy

The primary site of photography's monetisation has shifted away from the consumption of commodities and services and towards harvesting of data. This requires us to rethink many of the foundational assumptions about what photographs actually are. Or, rather, to recognise the disparity between what they look like to us (images), and what they mean to the corporations that profit from their public sharing (the basis for interactions that are surveyed as part of much larger data sets). If images continue to operate in the messy realm of individual and collective human experience, the algorithmic surveillance of our interactions with, through and around photography processes parts of that activity in statistical terms. It is for this reason that it presents some profound and fundamental challenges to artists and scholars. As Trevor Paglen (2014) neatly outlined, in an essay from 2014, a 'world of machine-seeing and invisible images' requires those 'interested in visual literacy… to spend some time learning and thinking about how machines see images through unhuman eyes, and train ourselves to see like them'.

If an engagement with photographs as images limits our understanding of photography as data, any parallel effort to attend to photography only as data risks reversing the current imbalance. Andrew Dewdney sums up the dilemma, observing that,

> this global condition of the algorithmic image continues to function within the field of representation, precisely because it remains as yet the humanly understandable surface of communication operating within common sense… we need an approach to understanding the interface between mathematical and cultural coding…to engage productively with the flat topology of the computer screen.
>
> *Dewdney 2016*

Here Dewdney highlights how the representational content of computational photography is not distinct, or somehow separate, from the operationalisation facilitated by its parallel condition as data. The algorithmic harvesting of digital information is based on the surveillance of activities promoted and encouraged by the content of different pictures. The locations we access, the people with whom we interact and the types of content we like, love or LOL about reveal different things about users, their preferences and their likely future behaviour. Images make data legible to humans as something other than data. The resulting interactions make humans legible to machines. Post-capitalist photography seeks to reverse that dynamic, using the content of images to develop forms of critical legibility through which the operations of the network become clearer to humans and thus build the knowledge of systems that enable them to be disrupted and undermined.

Some productive, if unlikely examples, can be identified within contemporary art projects that make use of networked images. Think, for instance, about Ed Fornieles' (2011) project *Dorm Daze*, a Facebook sitcom set at the University of California. It used information 'scalped' from the existing social media profiles of Berkeley students as inspiration for a cast of characters. Lives were performed on Facebook by the artist's friends according to a semi-scripted narrative. Over time, the characters deviated from stereotypes: a basketball star became involved with a violent drug gang; an emo and a jock explored the joys of sadomasochistic sex; a yoga lover became involved with Occupy and blew up a bank; and the father of a rich kid was discovered to be harvesting human organs in Cambodia. The requirement to present a single, stable identity is essential to the effective monetisation of our online behaviour. Even when the socially mediated world is understood to be inherently performative, it is the consistency of individuals' performances in particular contexts or situations that allow sociological patterns to be identified, specific demographics exposed and common desires to be exploited. Fornieles notes how his absurdist fiction 'pushed against the algorithmic environment within which it existed (the numbers, the data that feeds Facebook and other social media), turning analytics into a redundancy. Adverts and suggested friends become repurposed as scenery or props to a new fiction' (Petty 2020).

Art projects based on Street View photography by Jon Rafman and Michael Wolf were made by navigating Google Street View's image world through a curious and unstructured ramble, following no particular logic beyond an intuitive sense of what makes for an intriguing image. This is not the case with projects by Mishka Henner or Doug Rickard, which adopted a more structured approach that tied specific locations to economic dispossession. For Guy Debord (1956) the modern metropolis was a manifestation of capitalist domination, designed to ensure the smooth flow of workers and consumers in tightly scripted daily rituals. It was by allowing ourselves to 'be drawn by the attractions of the terrain and the encounters' during unplanned explorations that we might resist the circumscribed behaviours. Rafman and Wolf seek nothing other than an interesting image, applying the logic of the *dérive* to an exploration of Street View. Henner or Rickard maintain a more systematic approach: it is not hard to see how the data generated through the production of their series could find real-world instrumental uses.

Control today is exercised through the shadowy regimes of online surveillance. But photography links the activities surveyed to areas of the social and material world – from the identities we perform to the urban spaces we navigate – which are linked, in turn, to longer histories of activism and critique. There is much to be gained by reimagining these histories of resistance in the age of communicative capitalism.

Post-Capitalist Photography

For sociologists Luc Boltanski and Eve Chiapello (2005), networked capitalism was a response to the artistic critique levelled against mid-twentieth-century Fordism. This demanded greater autonomy, authentic experience, freedom from the monotony of mass-produced consumer culture and the drudgery of nine-to-five working. Capitalism responded by 'liberating' the workforce; fragmenting cultural and commodity production to meet the minutiae of personal preference; creating a flexitime world where success is gauged not through movement up definable hierarchies but by a capacity to move between projects and across global networks.

The shift has had a fundamental impact on subjectivity, on the processes through which we understand who we are. As citizens of what Boltanski and Chiapello call the 'projective city' – a social space 'founded on the *mediating* activity employed in the creation of networks' – we are unbound by traditional structures of family, church and profession. So we have to construct ourselves through the micro-identities we consume (Boltanski and Chiapello 2005, pp. 169–71). Work today is based on the relationships we develop and the connections we forge: on the extensiveness and effectiveness of our networks. As the distinction between labour and leisure becomes unclear, the identities we fashion play an integral role in what would once have been called our professional lives.

For cultural theorist Sarah Gram (2015), the effects come into the clearest focus in the figure of the teenage girl. Understood 'more as a concept than a biological necessity', the teenage girl is 'the central unit of late capitalism, the model citizen of commodity society'. Perceived to be useless as a worker under industrial capitalism, the teenage girl was made useful through the requirement to purchase. She became 'a worker whose primary labour is dedicated to looking a particular way rather than making a particular thing'. Her body was refashioned as a commodity, 'one which belongs to her and is her responsibility to maintain the value of'. Becoming 'a spectacle, a narcissist, a consumer are simply the criteria that have to be met to be legible under late-capitalism, particularly as young women'.

Jonas Larsen and Mette Sandbye are right to observe that the behaviour promoted by social media platforms has helped transform snapshot photography from a medium of memorialization to one of the communication (Larsen and Sandbye 2014, pp. xivi–xxxii). A cornerstone of domestic consumption has been radically redefined in the process. What was once called 'the Kodak moment' no longer suggests a private or semi-private memory, but an opportunity for more public forms of display. A former Kodak employee explained that the 'Kodak moment' was a way to scare people that unless they bought a little yellow box, their most important memories risked ruin…the ads were meant to make you cry, and if you cried, that meant you pay an extra 45 cents for a roll of film.

quoted Schumpenter 2012

Along with an increasing emphasis on youth, advertising today highlights the need to share our experiences, not just to preserve our memories. We are required to *show others* how much we care across social media platforms.

Prominent voices in photography's historical and theoretical analysis – from Jo Spence (1995) to Allan Sekula (1981) – have linked the medium's domestic applications to the logic of the nuclear family and, ultimately, private property. For the vast majority of people, experiences of making photographs involved the documentation of a relatively small part of the world to be preserved and viewed within the confines of people's homes. The integration of snapshot photography into social media platforms presents a complex reconfiguration of that dynamic. On the one hand, it has consolidated and intensified an earlier, individualistic logic: the bonds of family displaced by a perpetual performance of self that represents a kind of hyper-individualism. On the other hand, the sharing of photographs – now the default position of the image – may have displaced the private sphere of the family album with something altogether more public. A photograph uploaded onto Instagram or Facebook may be seen by relative few, and users are increasingly informed about the ways in which personalisation algorithms and filter bubbles limit who sees what. And yet the networked image is fundamentally different from the domestic snapshot in the *potential* audiences it implies: no longer a singular treasured object, but something that can and might be accessed and copied by multiple people simultaneously.

The implications of that change have been discussed by Jodi Dean (2016) in terms of 'selfie communism'. The individual self-fashioning encountered on social media platforms – a set of practices exemplified by the 'selfie' – could have, and could yet, lead to the formation of an unpredicted and paradoxical form of collective identification. Communicative capitalism has enlisted photography as a vital tool in a continual public performance of self in order to produce value for social media platforms' corporate shareholders. And yet shared social patterns have developed over time; rituals and common motifs that possess important, often unacknowledged, potential. The selfie, Dean suggests, should be understood not just in terms of individualism, but as 'a common form, a form that, insofar as it is inseparable from the practice of sharing selfies, has a collective subject'. The subject 'is the many participating in the common practice, the many imitating each other'. The act of freely sharing photographs, Dean argues, challenges the cult value of unique objects: 'Reproduction becomes inseparable from production…the image posted on Facebook can be on any number of screens at the same time, whether or not it even registers to anyone scrolling through'.

Dean knowingly updates Marx's adage that capitalism would produce its own gravediggers for a social media age. What originally referred to the unintended ways that bringing workers together in factories provided the space within which their collective interests could be recognised is recast in terms of a 'secondary visuality' linked to the formation of collective identification – a sense of being that we share and a sense of our shared being – that emerges as an unintended consequence of our unique selves being performed publicly, online and *en masse* (Dean 2016). It is in this sense that 'selfie communism' can be counted as a specific iteration of what Kelsey and Stimson (2008) call the 'sociality of photography': a photographic politics based on connections and flow not just the semiotics of the image.

In the context of my argument, what matters most about Dean's thesis is a reconfiguration of our relationship to culture, to each other, and, particularly, to work. A platform such as Instagram serves diverse purposes, some of them linked directly or indirectly to professionalised creativity. Yet the overriding experience of scrolling through its endless image flow is rooted in the abundance of creative content being produced, shared and viewed for free. As Dean suggests, the political potential of such a culture emerges only when it is coupled with an understanding of the platform's extraordinary economic value (in August 2021, Facebook Inc. was valued at $1.21 trillion). It is here we confront the fundamental question of how characteristics true of our experiences of networked images might be extended to ownership of the network itself and, particularly, distribution of the wealth it generates.

Photography after Capitalism

Photography's economic condition has always taken shape in relation to broader formations, whether Fordist mass production and consumer culture at the middle of the previous century or communicative capitalism today. So alternatives will also be forged in relation to larger politico-economic projects. Therein lies the utility of post-capitalism. One possibility involves the migration of online photographic lives, from the spaces owned by corporate giants onto decentralised, user-owned platforms. Egalitarian, cooperative and ethical alternatives hold much in common with the types of post-capitalist politics outlined by the feminist economists J.K. Gibson-Graham (1996, 2006), for whom it is vital to challenge the 'capitalocentrism' that has too often marked critiques of neoliberalism among the despondent left. Seeking out alternative economies in the here and now signals what we are capable of achieving (Gibson-Graham 1996, 2006). The rapid ascent of Stocksy in the stock image marketplace – an 'artist-owned cooperative' that ensures 'empowered shareholder artists receive fair pay' – as a rival to Getty Images signals the potential of that move. Getty's use of digital platforms to 'democratise' the provision of stock imagery has radically de-professionalised the field, opening up supply in order to drive down the royalty payments it is compelled to pay photographers. Stocksy, by contrast, extends democratic principles to the ownership and management of the company (Scholz and Schneider 2017).

That alternative social media platforms such as Diaspora, PixelFed or Friendica have hitherto failed to mount a significant challenge to giants such as Instagram or Facebook signals the considerable barriers faced, particularly what Srnicek calls 'network effects': the 'cycle whereby more users beget more users, which leads to platforms having a natural tendency towards monopolisation'. Even 'if all its software were made open source, a platform like Facebook would still have the weight of its existing data, network effects, and financial resources to fight off any co-op rival' (Srnicek 2017, p. 125). The struggle to achieve prominence may also indicate that our Faustian pact with multi-nationals may not actually bother most users, who accept surveillance and advertising as the price that had to be paid for 'free' access.

The idea that image-sharing platforms be brought under collective ownership has real appeal. The same is true of Google Earth and Street View. Were this taken as an opportunity to remove surveillance and advertising, then social media would shed the characteristics that tie it to regimes of 24/7 production and become something closer to the participatory media outlet and public playground it purports to be. If data extraction continued, however, the extraordinary profitability of such sites could be directed towards creating better conditions for those workers who currently lack them, and towards a variety of public projects, in the same way that social housing represents a just and common sensical alternative to the private rental market. That surveillance could have suspect applications signals the importance of building democratic accountability into future plans; it is extraordinary that we should find any similar accountability almost entirely absent from the current corporatised model.

Refashioning platforms as publicly owned utilities would initially prove costly, and exceptionally complex, given the multi-national operations of platform giants and the national territories served by most public institutions. It is more straightforward to imagine steps being taken to break up monopolies and to ensure that a greater proportion of the wealth generated by privately owned platforms is reclaimed. Jaron Lanier once envisaged a system of micro-payments to remunerate individual users based on their specific contributions (Naughton 2013). I think it simpler, and more politically sentient, to increase taxation on platforms, harnessed to fund the public services users require, along with the provision of a UBI.

UBI can be seen as a pragmatic measure, required to ward off the effects of automation and an economy in which extraordinary levels of wealth remain concentrated in the hands of small elites. In the context of post-capitalist thought, it possesses an important utopian dimension, creating mechanisms through which people would be liberated from waged labour and offered greater choice about how to use their time. For economists Philippe van Parijs and Yannick Vanderborght (2017), UBI enables people to say yes to the meaningful jobs that do not necessarily pay so well, and no to the meaningless jobs that they do only for money.

For Mariana Mazzucato, UBI is most strategically described as a 'citizen's dividend', co-opting the language of the boardroom to shape redistribution as a rational reflection of common ownership. Mazzucato's primary concern is the numerous ways in which corporate wealth production relies on public investment: from government-funded research to public infrastructure. It is perverse that 'we socialize risks but privatize rewards' (Mazzucato 2020). Those sentiments take on a particular character in the context of post-capitalist photography. If our social lives as images, and the social lives of images, are harnessed as the basis for profitability, part of that income should be redistributed to sustain the social world without which profits would not be possible. Funded through taxation on the most wealthy in society – a category that includes platform owners – UBI has the rare ability to answer the artistic and social critiques to which capitalism has historically been subject. Structural inequities are challenged through the redistribution of wealth, while individuals are provided with greater autonomy to decide what to do with their time.

Distinctions between these options have too often been framed as a binary. Supporters of cooperatives perceive structural change as a deferral of action, while proponents of larger transformations regard community ventures as small-scale distractions from the more fundamental activity required. But the relationship between the two need not be understood as a stark choice: platform co-ops today *or* UBI tomorrow. One can provide a stepping-stone towards the other, creating spaces where collective subjectivities are nurtured, developed and maintained. It remains imperative therefore that means are not mistaken for ends.

How can alternatives be brought into being? The creation of a post-work world is a long-term project, a matter of 'decades rather than years, cultural shifts rather than electoral cycles' (Srnicek and Williams 2016, p. 107). Williams and Srnicek envisage a broad 'ecology of organisations' operating 'in a more or less co-ordinated way, to carry out the division of labour necessary for political change'. Popular movements nurture collective subjectivities and affective bonds. Foundations and journalists change mainstream media narratives. Intellectual organisations such as think tanks, educational institutions and more informal consciousness raising-bodies enable ideas to be produced and shared. Labour organisations, retuned to the reality of contemporary work and more closely engaged with wider communities, take struggles into the workplace. The state can legislate, redistribute wealth and finance alternative models. A technical understanding of machines becomes 'vital to understanding how to disrupt them' (Srnicek and Williams 2016, pp. 162–8).

A project such as *Wages for Facebook* – essentially a conceptual gesture, via which artist Laurel Ptak demanded payment for time spent on social media – signals the extent to which the formulation of means and ends can sometimes serve each other. Described not as 'a thing', but 'a political perspective', the campaign names a possibility in order to build awareness. It is in that sense closer to historical struggles to build consciousness about unwaged domestic work than to industrial action. Something similar can be suggested about the possibilities rehearsed above, which signal the profound inequities built into the corporatized platform model by identifying alternatives.

Raising awareness matters as a necessary first step, but at some point, it is necessary for objectives to be identified and think carefully about the strategies via which they can be pursued. *Collectivise Facebook*, a 2021 project by artist Jonas Staal

FIGURE 3.1 Jonas Staal and Jan Fermon, *Collectivise Facebook: A Pre-Trial,* 2021. Theatre Rotterdam, produced by HAU Hebbel am Ufer, Berlin. Photo: Ruben Hamelink. Courtesy Jonas Staal

and lawyer Jan Fermon, represents an interesting example in such a context (see Figure 3.1). Where *Wages for Facebook* was essentially a thought experiment via which alternative horizons were brought into view, *Collectivise Facebook* attempts to exploit a legal apparatus to realise its goals. A collective action lawsuit submitted to United Nations Human Rights Council in Geneva aims 'to force legal recognition of Facebook as a public domain that should be under ownership and control of its users: Facebook must be collectivised'. Even when the project is understood as being essentially performative, it nevertheless directs attention towards the available mechanisms via which goals can be pursued.

Global Challenges

Writing in 2013, Jaron Lanier observed that, at

> the height of its power, the photography company Kodak employed 140,000 people and was worth $28 billion. But today Kodak is bankrupt, and the new face of digital photography has become Instagram. When Instagram was sold to Facebook for a billion, it employed only 13 people. Where did all those jobs disappear?
>
> *Lanier 2013, xii*

It is an important question, with several potential answers. Lanier's primary concern is the move towards image 'prosumption' with which I began this chapter. He is much less interested in the specific roles once performed by workers at Kodak's factory or the related forms of labour that have emerged in the context of the networked image economy. The expanded view of digital labour, developed by scholars such as Christian Fuchs (2014) and Enda Brophy and Greig de Peuter, provides a more substantive answer, highlighting the 'circuits of exploitation' that link the making and sharing of digital content to processes of 'extraction, assembly and design through mobile-work, support work and e-waste' (Brophy and de Peuter 2014, pp. 60–1).

Where the disavowal of work in the context of image prosumption is linked to the ambiguous relationship of labour and leisure, and the challenges of conceptualising data extraction as a novel form of exploitation, here, it is a matter of work's more literal non-visibility, concealed behind black box technologies and geographical partitions. Shining light on that work is an important step towards the identification of a common operational logic and the recognition of shared interests. The answer to Lanier's questions lie as much in casualisation, outsourcing and globalisation as they do in technological change.

The vast majority of cameras sold today are integrated into smartphones. The devastating conditions under which that labour is performed are highlighted by *Phone Story* (2011), a gaming app for smartphones produced by artist-activist Molleindustria. It adopts the format of a series of low-fi computer mini-games that require players to assist in the production, consumption and disposal of i-Phones: manipulating armed guards to make workers mine for coltan (a central component in electrical capacitors); catching suicidal workers in the Foxconn factory where Apple products are made; coercing consumers into the Apple store to buy new products; and disposing of e-waste in Pakistan. This expanded perspective on a global image economy brings new histories of photography into view: no longer only a matter of photographers, or even camera manufacturers, but also silver mining in Ontario and the disposal of toxic waste materials in the rivers around the Kodak factory in Rochester (Angus 2021).

Social media content moderators are required to monitor the billions of images uploaded to platforms every day to remove offensive content. Sarah Roberts (2017) describes a variety of workplace organisations stretching around the globe. While some of the work takes place in large call-centre environments in the Philippines and India, it is also farmed out as piecemeal micro-work in the global gig economy. The work of making decisions about that content is complex, particularly when it is outsourced to other countries. Workers must 'become steeped in the racist, homophobic, and misogynist tropes and language of another culture' and are thus often required to put aside personal belief systems and morality. While platforms that rely on user-generated content face intense commercial pressures to remove it, some degree of offensive content is important to many sites because it drives up attention, clicks and traffic. As a result, workers find themselves in a paradoxical role, in which they must balance the site's desire to attract users and participants to its platform – the company's profit motive – with demands for brand protection, the limits of user tolerance for disturbing material, and the site rules and guidelines.

Roberts 2017

Dark Content (2015), a project by artists Franco and Eva Mattes, used avatars and voice software to animate accounts provided by real content moderators. One moderator, who lived in their car and used Wi-Fi connections in fast-food restaurants, recalled being

> in McDonald's reviewing images that were fairly dense with hard-core porn…About three hours in, I got up to use the restroom and looked behind me. There was a family sitting there, able to see everything I was doing. Apparently there was a door I hadn't noticed and people were coming and going the whole time.

Another worker tried to avoid the 'requesters' who ask them to look at suggestive images of children or people engaging in sexual activity with animals.

The profitable circulation of imagery is increasingly reliant on AI-based systems of image recognition. Linking visual forms to specific concepts without requiring specific images to be labelled with keywords helps to maximise the efficiency of internet search engines, for example. Yet machine-seeing is still reliant on input from human workers. Invitations to tag the faces of friends on Facebook help train algorithms to recognise different faces. Google's 'reCAPTCHA' authentication tool, which asks humans to verify their humanness by deciphering a jumbled text or image and typing details into a computer, has previously used sections of photographs from Street View. When typing in house numbers, or identifying cars, trees, and street signs, the requirement that humans demonstrate they are not machines helped train machines to see like humans. The results find applications in Google's driverless cars, which deploy a combination of GPS, censors and a rooftop LIDAR, which uses lasers to measure the distance of objects, and a standard camera directed through the windscreen, which 'looks' for nearby hazards, reads road signs and detects traffic lights.

The Yahoo Flickr Creative Commons 100 Million, Dataset or YFCC100M – a large and freely usable dataset used in machine learning-was compiled – was compiled using photographs uploaded onto Flickr with a Creative Commons license. Users can buy expansion packs to supplement the dataset, which include 'autotags' identifying the 'presence of visual concepts, such as people, animals, objects, events and architecture'. The utility of the Flickr photographs lay not only in the abundance of images, but in the existing tags and categories provided for photographs by the platform's users. Paid roles are usually outsourced to casual labourers via platforms like Amazon Mechanical Turk, 'a digital labour market where workers from across the world and around the clock browse, choose, and complete human intelligence tasks that are designed by corporate or individual contractors' (Aytes 2012, p. 80). The platform exemplifies the logic of low-paid, outsourced, casual working that defines so much of neoliberalism.

This itemisation of specific roles is an important task, when so much of the networked images' economic condition remains unreported and unseen. And yet, in the words of Alexander Toscano and Jeff Tinkle, 'detection and discovery fall short'. While 'exposure makes for important political work, it has to be linked to systemic concerns' (Toscano and Tinkle 2008, p. 68). The true significance of such a move lies, instead, in thinking carefully about the relationships between a system's constituent parts. To return again to Lanier's question, the assault on stable employment that has characterised the transition from an industrial to digital economy cannot be explained via technological change alone. As sociologist Saskia Sassen (2014) has signalled, a common operational logic binds experiences such as these to each other, and to the harvesting of users' data: the merciless minimisation of labour costs; a ruthless maximization of profit; the deliberate evasion of social responsibility by hoarding funds offshore.

Speaking in 2019, Jodi Dean highlighted those ways in which accelerationist fantasies of societies liberated by automation, and living on UBI, often ignore the manual production and brute extraction on which machines rely. The international solidarity and managed economies of communism, Dean argues, dissolve national boundaries and so bring different choices into view: do I want a smartphone so much that I would be prepared to spend part of my working life in coltan mines? Do I desire an internet free of brutal images to the extent that I am prepared to do my time as a content moderator?. The promise of post-capitalist photography is not a simple argument for more and more images available for free as part of participatory cultures but also a way of accessing more profound questions about egalitarianism and its implications. It names the need for global solidarity, along with the need not to neglect the former blue-collar workers, pushed to the 'systemic edge' by outsourcing and automation, 'so dispirited that they stopped looking for work altogether'. Former factory towns such as Rochester, historic home of Kodak, provided a strategic and symbolic focus for Trump and his message: providing an irrevocable demonstration of the ruthlessness with which alienation will be exploited by the coalitions that continue to gather on the political and economic right.

Post-capitalist photography uses the ubiquity of the networked image to map the inequities of a neoliberal world order, and to situate cognitive and manual labour in direct relation to each other. At the same time, it harnesses the world that networked photography brings into view as opportunity to confront the asymmetry that currently defines photography's interactions with the network. An indication of all that is right and wrong with the image worlds of today, and an image of what the world could be tomorrow, post-capitalist photography describes a powerful sense of our shared interests, along with the powerful barriers that currently stand between us and their realisation.

Bibliography

Andrejevic, Mark. 2011. "Estrangement 2.0", *World Picture* 6 (Winter 2011): www.worldpicturejournal.com/WP_6/Andrejevic.html (accessed 3 February 2020).

Angus, Siobhan. 2021. "Mining the History of Photography." In *Capitalism and the Camera: Essays on Photography and Extraction*, edited by Kevin Coleman and Daniel James. London and New York: Verson, 55–74.

Aytes, Ayhan. 2021. "Return of the Crowds: Mechanical Turk and Neoliberal States of Exception." In *Digital Labour: The Internet as Factory and Playground*. edited by Trebor Scholz. New York and London: Routledge, 2012, pp 79–97.

Batchen, Geoffrey. 2000. "Vernacular Photographies", *History of Photography* 24 (3): 262–271.

Beech, Dave. 2019. *Art and Postcapitalism Aesthetic Labour, Automation and Value Production* London: Pluto.

Benjamin, Walter. 2008. *The Work of Art in the Age of Mechanical Reproduction*. London: Penguin.

Boltanski, Luc and Eve Chiapello. 2005. *The New Spirit of Capitalism*, trans. Gregory Elliot London and New York: Verso, 2005.

Brophy, Enda and Greig de Peuter. 2014. "Labours of Mobility: Communicative Capitalism and the Smartphone Cybertariat." In *Theories of the Mobile Internet: Materialities and Imaginaries*. edited by Andrew Herman et al., London and New York: Routledge, 2014, pp 60–86.

Crary, Jonathan. 2013. *24/7: Late Capitalism and the Ends of Sleep*. London and New York: Verso.

Dean, Jodi. 2016. "Images without Viewers: Selfie Communism", *Still Searching: Fotomuseum Winterthur* (1 February 2016): www.fotomuseum.ch/en/explore/still-searching/articles/26420_images_without_viewers_selfie_communism (accessed 3 February 2020).

Dean, Jodi. 2019. "Communism or Feudalism", *Sonic Acts Festival* (23 February 2019): www.youtube.com/watch?v=XGq3hdWEe10 (accessed 3 February 2020).

Debord, Guy. 1956. "Theory of the Dérive", *Les Lèvres Nues #9* (November 1956). trans. Kenn Knabb, published on *Situationist International Online*: www.cddc.vt.edu/sionline/si/theory.html (accessed 3 February 2020).

Dewdney, Andrew. 2016. "Co-Creating with Networks: A Reply to 'What is 21st-Century Photography?'", *The Photographers' Gallery Blog* (4 January 2016): http://thephotographersgalleryblog.org.uk/2016/01/04/co-creating-in-the-networks-a-reply-to-what-is-21st-century-photography/ (accessed 3 February 2020).

Farocki, Harun. 2004. "Phantom Images." Public 29: 12–22.

Fisher, Mark. 2021. *Postcapitalist Desire: The Final Lectures*. London: Repeater, 2021.

Fisher, Mark. 2016. "Baroque Sunbusts." In *Rave and its Influence on Art and Culture*, edited by Nav Haq. London: Blackdog, pp 39–48.

Fuchs, Christian. 2014. *Digital Labour and Karl Marx*. London and New York: Routledge.

Gibson-Graham, J.K. 1996. *The End of Capitalism (As We Knew It)*. Minneapolis, MN: University of Minnesota Press.

Gibson-Graham, J.K. 2006. *A Post-Capitalist Politics*, Minneapolis, MN: University of Minnesota Press.

Gram, Sarah. 2013. "The Young Girl and the Selfie", *Textual Relations* (1 March 2013): http://text-relations.blogspot.com/2013/03/the-young-girl-and-selfie.html (accessed 3 February 2020).

Hand, Martin. 2012. *Ubiquitous Photography*. Cambridge: Polity.

Hester, Helen. 2017. "Demand the Future: Beyond Capitalism beyond Work", *Demand the Impossible* (22 March 2017): www.youtube.com/watch?v=-VlaGnX1eyc (accessed 3 February 2020).

Hoelzl, Ingrid and Remi Marie. 2014. "Google Street View: Navigating the Operative Image", *Visual Studies* 29 (3): 267.

hooks, bell. 2003. "In Our Glory: Photography and Black Life." In *The Photography Reader*, edited by Liz Wells. London and New York: Routledge, 387–94.

Jameson, Fredric. 1987. "Cognitive Mapping." In *Marxism and the Interpretation of Culture*, edited by Carry Nelson and Lawrence Grossberg. Champaign, IL: University of Illinois, 347–60.

Kelsey, Robin and Blake Stimson. 2008. "Introduction: Photography's Double Index (A Short History in Three Parts)." In *The Meaning of Photography*, edited by Robin Kelsey and Blake Stimson. Williamstown, MA: Sterling and Francine Clark Art Institute, xxiii–xxiv.

Lanier, Jaron. 2013. *Who Owns the Future?*. London and New York: Penguin.

Larsen, Jonas and Mette Sandbye. 2014. "Introduction." In *Digital Snaps: The New Face of Photography*, edited by Jonas Larsen and Mette Sandbye. London and New York: I.B. Tauris, pp xv–xxxii.

Lovink, Geert. 2011. *Networks without a Cause: A Critique of Social Media*. Cambridge: Polity.

Lovink, Geert and Marianne Rasch. (eds.). 2013. *Unlike Us Reader: Social Media Monopolies and their Alternatives*. Amsterdam: Institute of Network Cultures.

Mason, Paul. 2016. *Postcapitalism: A Guide to Our Future*. London and New York: Penguin.

Mazzucato, Marianna. 2020. "We Socialize Bailouts. We Should Socialize Successes, too", *UCL News* (2020): www.ucl.ac.uk/news/2020/jul/opinion-we-socialize-bailouts-we-should-socialize-successes-too.

Moholy-Nagy, Lazlo. 1987. *Production/Reproduction*. London: Thames and Hudson.

Naughton, John. 2013. "Jaron Lanier: The Digital Pioneer Who Became a Web Rebel", *The Observer* (17 March 2013): www.theguardian.com/technology/2013/mar/17/jaron-lanier-digital-pioneer-rebel (accessed 3 February 2020).

Paglen, Trevor. 2014. "Safety in Numbers", *Frieze* (12 March 2014): https://frieze.com/article/safety-numbers (accessed 3 February 2020).

Petty, Felix. 2016. "Performing Ourselves: Revisiting Dorm Daze with Ed Fornieles", *Dazed* (23 March 2016): https://i-d.vice.com/en_au/article/ywvkw5/performing-ourselves-revisiting-dorm-daze-with-ed-fornieles (accessed 3 February 2020).

Roberts, Sarah T. 2017. "Behind the Screen: Commercial Content Moderation", *The Illusion of Volition*. https://illusionofvolition.com/behind-the-screen/ (accessed 3 February 2020).

Sassen, Saskia. 2014. *Expulsions: Brutality and Complexity in the Global Economy*, Cambridge, MA and London: Harvard University Press.

Schumpenter, Mike. 2012. "How Fujifilm Survived", *The Economist* (18 January 2012): www.economist.com/schumpeter/2012/01/18/sharper-focus (accessed 3 February 2020).

Scholz, Trebor. 2012. "Introduction: Why Does Digital Labour Matter Now?." In *Digital Labour: The Internet as Factory and Playground*, edited by Trebor Scholz. New York and London: Routledge, pp 1–9.

Scholz, Trebor and Nathan Schneider. 2017. *Ours to Hack and to Own: The Rise of Platform Cooperatism, A New Vision for a Fairer Future and a Fairer Internet*. New York and London: OR Books.

Sekula, Allen. 1981. "The Traffic in Photographs", *Art Journal* 41 (1): 15–21.

Sluis, Katrina. 2014. "Authorship, Collaboration, Computation", *Photoworks Annual*, 21, 151–9.

Sluis, Katrina. 2018. Interviewed by Lewis Bush, *1000 Words* (28 March 2018): www.1000wordsmag.com/katrina-sluis/ (accessed 3 February 2020).

Srnicek, Nick and Alex Williams. 2016. *Inventing the Future: Postcapitalism and a World without Work*. London and New York: Verso.

Srnicek, Nick. 2017. *Platform Capitalism*. Cambridge: Polity.

Spence, Jo. 1995. *Cultural Sniping: The Art of Transgression*. London and New York: Routledge.

Toscano, Alberto and Jeff Tinkle. 2015. *Cartographies of the Absolute*. Winchester and Washington: Zero.

van Parijs, Philippe and Yannick Vanderborght. 2017. *Basic Income: A Radical Proposal for a Free Society and a Sane Economy*. Cambridge, MA and London: Harvard University Press.

PART II

Computation, Software, Learning

4

THE COMPUTER VISION LAB

The Epistemic Configuration of Machine Vision

Nicolas Malevé

Computer vision,[1] the discipline that aims to emulate human visual abilities through software, has become a utility that every piece of code can make use of in order to parse visual data. Through its inclusion in many popular products such as cell phones, social media platforms, car navigation systems or satellite imaging, it demonstrates the ability of machines to perform cognitive tasks such as the identification of objects, people or land. Computer vision equips digital devices with visual intelligence on demand. More than a niche product, computer vision has become infrastructural as it is now deployed at planetary scale as a service, as a code package or embedded in devices.

The ubiquity of computer vision underscores the importance of understanding precisely the relation between machine vision and the forms of knowledge that give rise to it. To do this, this chapter will approach computer vision in its formation rather than in the stabilised forms of software products. The objective of this text is to raise the question of the epistemic configuration of computer vision: the different bodies of knowledge that are mobilised, the actors whose contribution are acknowledged (or not) and the entanglement of institutions of research with environments of industrial production. In these pages, I focus on the computer vision lab, a space where computer scientists experiment with human vision in order to make it tractable to algorithms. I will analyse the terms on which this translation is made, and in particular how logistics, labour conditions and theories inform each other. From this perspective the chapter approaches the computer vision lab as the site where computer scientists engage experimentally with cognitive psychology in a context where the laboratory and the industrial environment of production are increasingly entangled.

To start unpacking these questions, I will turn to the work of a particular group of researchers whose central figure is Fei-Fei Li.[2] Her trajectory connects academic work to industrial development. Li, who has been in turn director of the Stanford Artificial Intelligence (AI) Lab, vice president at Google Cloud and co-director

DOI: 10.4324/9781003095019-7

of Stanford Institute for Human-Centered AI, is best known for the creation of ImageNet, a dataset of 14 million images that contributed significantly to the development of machine vision (Deng et al. 2009). Analysing Li's work, I will reflect on the different forms of knowledge at play in the development of machine vision and move beyond a narrative of computer vision that tends to limit itself to code and mathematics. As a point of departure, I will concentrate on one of its more prosaic aspects: the training process of the algorithm.

The developments achieved by computer vision algorithms are obtained through an elaborate translation of the ways in which humans see, interpret and produce images. To emulate these cognitive abilities, computer vision algorithms make heavy use of collections of images called datasets. A dataset in computer vision is a curated set of digital photographs[3] that developers use to test, train and evaluate the performance of their algorithms. The algorithm is said to learn from the examples contained in the dataset. In the current stage of development of computer vision, the size of datasets is a key factor. These collections of images must represent the variety and regularities of the visual world. Scale is therefore a crucial factor in dataset's creation. In a conversation with Google engineers, Fei-Fei Li, ImageNet's creator, discusses the challenges she faced when assembling such a large collection of photographs, and the scale at which the work of selection and verification needs to be performed. For Li, a tipping point had been reached when, having downloaded 10,000 images for each of the 40,000 categories of her dataset, each image needed to be verified. Li had used the Google Images search engine to fetch the photos through keyword searches. But the results were far from accurate. As a result, she needed human collaborators to look at them and filter out erroneous results. But how can this be achieved? As Li explains:

> … let's just say a person just looks at 2 images per second and doesn't eat and sleep and so on. And if we do this computation to construct the entire ImageNet, it will take 19 human years. So I ask my graduate student, do you want to do this? (Laughter) He said no. I need to graduate.
>
> *Fei-Fei 2011*

The dataset's scale exceeds the limits of one person who would have to dedicate 19 years without sleeping or eating. Asking her student to do the work, she is faced with a resounding no.

Beyond its informal character, her remark about the grad student introduces various elements that are of interest in unpacking the different epistemic configuration of machine vision. The first concerns the labour required to train machines to see. Current machine vision relies on powerful algorithms which are able to generalise the relevant features extracted from a set of samples. Scale is of the essence as large and diverse datasets increase the precision of the algorithms trained with them. The production of these datasets is human labour-intensive. It presupposes that photographs, the product of creative labour of many web users, are available to the scientists by millions. Further, in her account Li explains that she initially thought the work of verification could be performed for free or cheaply by people at an early stage of their curriculum, graduate students. Yet this proves impossible due to the immensity of the task: students have classes to attend and exams to pass.

Li's anecdote implies that verifying images, although a crucial part of the dataset creation, is not an official part of the training of computer scientists and therefore should not be expected from a graduate student. The graduate student does not refuse because, as a person living in a body, they have to eat and sleep, but because as a stakeholder in the education system they have to graduate. The story addresses both the non-human scale of the dataset and the implied cruelty to expect someone to adjust to this scale, as well as touching upon different epistemic configurations. The story interlaces different meanings of positions, grades and scales defining what counts as labour, as learning and as human. Further, the anecdote underlines the limits of what can be handled within a knowledge institution such as the university. It talks about what can be taken for granted (the availability of images and therefore creative work) and what will need to be bargained: the cognitive labour of "verification." Finally, the story refers to a tempo, an act of classification that can be performed in half a second ("two images per second"). For dataset creators, it seems that classification happens in a glance. The anecdote raises the question of the kind of subject who can perform classification at this speed without any consideration for their body condition during 19 years. A subject only engaged in an act of perception and who doesn't have to graduate. A viewer able to embody the speeds and duration implied by the scale of the dataset. Who is this person, and more importantly how and where is he constructed?

Experimental Practice in the Computer Vision Lab

Before computer scientists start coding, a series of objects come to them readily legible. The existence and the definition of a collection of concepts as disparate as vision, viewers, photographs and objects of all kinds are already assumed. In considering the epistemic configuration that gives rise to them, I will first concentrate on the informal group of researchers studying vision at Caltech and who initiated the creation of many hallmark datasets for machine vision. In the conceptual universe that imbues their research, there is a privileged level at which the world welcomes perceiving subjects and where these notions offer a stable conceptual ground. Here, an apple, a dog, a microwave are concepts that can be detected and labelled without resorting to sophisticated reasoning, and there is a level at which the world can be perceived and shared consensually (Fei-Fei et al. 2007; Griffin et al. 2007; Perona 2010). These are basic category objects available through common sense, and to establish what is basic, common and available through the senses requires the development of a material and conceptual framework. Li and colleagues, in parallel to their nascent work on datasets, are engaged with cognitive psychology and in particular psychophysics, a branch of psychology that studies the relation between the quantitative aspect of a stimulus and the probability of a particular judgement (Read 2015). These disciplines translate objects and relations sedimented through culture and history into a stable and operational form. Attending to the epistemic configuration that informs machine vision requires an inquiry into the role cognitive psychology plays in it. In what follows, I contend that cognitive psychology unifies and stabilises objects that circulate through software applications that emulate human cognition. Through their engagement with cognitive psychology, computer scientists seek a solution that collapses into the same plane the biological, cultural and technological dimensions of perception. I position the "computer vision lab" as the site where computer scientists accomplish this task. More than a physical building where scientists from different disciplines meet and work, here I am using the term "computer vision lab" to refer to a complex web of relationships where a set of methodological guidelines and constraints, embedded ontologies, image practices and labour conditions are invented, recycled and negotiated. And crucially, as a site where the border between knowledge institution and the industry is redrawn.

Vision at Speed

A good place to start examining the objects and concepts emerging from the computer vision lab is the act of vision. In this environment, because time is understood as a crucial intervening factor in visual perception, visual perception is construed as an act that unfolds in a particular temporal context whose mapping has been the object of an impressive amount of work (Fabre-Thorpe et al. 2001; Marr 1982). Vision is understood as a crossing of successive temporal thresholds, each connected with a particular cognitive function. In the computer vision lab, the early stage or first moments of visual perception are given specific attention. Early vision is treated as a black box unavailable to introspection: it resists cognitive influence and functions without the awareness of viewers. Early vision is active in the formation stage of visual cognition and it represents the delay needed for a first wave of information to pass through the visual system (Fabre-Thorpe et al. 2001, p. 171). This emphasis on early vision has its roots in Gestalt theory and neuropsychology, which have long considered the eye as an organ whose function is not limited to the mere transmission of the stimulus it receives from the outside world. The eye already interprets, it already 'speaks to the brain in a structured manner' (Lettvin et al. 1959 p. 1950). It already abstracts since 'a percept is a categorical shape rather than a mechanically faithful recording of a particular stimulus' (Arnheim 1969, p. 81). Cognitive science offers to computer scientists a model of early vision that pertains to visual cognition in its incipience. The pioneering work of physiologist Benjamin Libet posits that a short temporal interval separates bodily responses to stimuli from the moment where a decision is becoming conscious (Leys 2011), in which there is a missing half second between the initial neural reaction and the consciousness of an event. Crucially, cognitive psychology proposes a model of decision-making based on visual stimulus that contrasts with more traditional models of deliberation happening only in the mind. In the words of cognitive psychologist Zenon Pylyshyn (1999), vision is a discontinuous process where the first half second is encapsulated from awareness. Vision is therefore positioned in cognitive psychology as the moment where awareness enters the process of visual perception: a key moment where bottom-up (nervous processing of the stimulus) and top-down (mindful deliberation) forms of information processing start completing – and competing with – each other. In his meditation on Libet's work, Brian Massumi writes '[t]he conclusion has to be that the elementary unit of thought is already a complex duration before it is a discrete perception or cognition' (2002, p. 195).

Interest in early vision goes hand in hand with a reassessment of the importance of higher forms of cognition like reasoning and symbolic manipulation. Katherine Hayles (2020), in her detailed review of neuroscientific theories proposes the term nonconscious cognition to describe the permanent activity of embodied cognition, which never reaches consciousness but makes possible all the micro-decisions humans beings are taking continuously. Consciousness is treated like a retrospective narrative trying to maintain an illusion of coherence for a subject. As Hayles (2020, p. 10) explains, the slow uptake of consciousness represents a weakness as it is out of sync with the pace at which stimuli impinge on the nervous system. Its slowness represents a cost. As Massumi puts it, awareness is 'backdated' so that thought might 'coincide with itself' (Massumi 2002, p. 195).

This gives us a first idea of the conceptual background against which early vision is modelled and approached in the computer vision lab. It is particularly significant to the collaboration between neuroscientist Christof Koch, computer vision expert Pietro Perona and Fei-Fei Li in the first decade of the century who studied together the problem of scene understanding. The term scene understanding was used to describe the ability of human subjects to make sense of a scene in a glance, and became related to a particular model of visual cognition. Several experiments by Li and Perona concentrated on the reformulation of the cognitive model of attention which underlies scene understanding. These experiments were remarkable not only for the ideas that were tested but also for the experimental apparatus the researchers deployed. Because early vision is unavailable to introspection, it is only experimentally tractable using a technology that can control behaviour at a micro-temporal level. It must be studied from the outside and opened up to inquiry by experimental tools, which leads cognitive psychologists to design devices to decompose the course of perception.

Techniques such as rapid serial visual presentation (RSVP) illustrate this point. RSVP is a technique for consuming media rapidly by aligning it within the foveal region of the retina and advancing between items quickly. In many of these experiments Fei-Fei Li and her colleagues made heavy use of the RSVP technique trying to formulate scene understanding in terms of immediacy. In the experiment 'What Do We Perceive in a Glance of a Real-World scene?' (Fei-Fei et al. 2007), 22 students from the California Institute of Technology were cued repeatedly before an image for a very short time, from 27 to 500 milliseconds and asked to describe what they saw. The researchers demonstrated that certain levels of descriptions correlate with specific micro-temporal thresholds: at 27 milliseconds, subjects could describe amorphous blobs of colours and basic shapes in an image stimulus. At 200 milliseconds, they perceived generic objects such as tables and chairs. And at 500 milliseconds, they were able to itemise objects and people as well as describe their relations. In another experiment, 'Rapid Natural Scene Categorization in the Near Absence of Attention' (Fei-Fei et al. 2002), the researchers studied what subjects could capture from objects presented to their peripheral attention when concentrating on a different task. Subjects were asked to recognise letters shown for a brief moment while natural scene images appeared on the side of the screen. Even if their attention was directed to the task of recognising letters, subjects displayed an impressive ability to categorise the images appearing concurrently on the screen. This led the experimentalists to conclude that 'that there is little or no attentional cost in rapid visual categorization of complex, natural images' (Fei-Fei et al. 2002, p. 9596). It was through such experiments that researchers broke with a cognitive model of scene understanding happening in the mind in the form of an analytical reconstruction of logically connected parts. Instead, they moved to the concept of the *gist*, a general intuition of the elements pertaining to a scene captured in a glance, leveraging for that purpose the accuracy of micro-temporal measuring devices and techniques such as RSVP. As a consequence, they produced a series of heuristics such as the number of fixations per second necessary to make sense of a visual stimulus. They also devised the experimental conditions required to enact a subject of vision who is optimised to respond to rapid stimuli.

Classification

As we have seen, the encounter between computer vision and cognitive psychology produces a concept of vision unfolding through a micro-temporal timeline and concentrates on the glance as a privileged moment of perception. This opens the question of the nature of such percept. What does a gist consist of and how can it be described? Psychologist Eleanor Rosch's theory of classification gave to Li and colleagues important elements to answer this question.[4] Rosch's theory of categories marks a break in a tradition that framed classification as a problem of pure logic rather than a psychological problem (Rosch 1978). In her view, psychologists have inherited a model of classification that privileges logical consistency over cue validity. A class for a logician is a set of elements sharing the same definition. According to logicians, in a class, all elements are being given equal membership. In contrast, Rosch argues that categories as we use them in our daily activities are first and foremost analogue rather than logical, and categories have gradients of membership. A class can therefore be empirically understood by the identification of elements that are typical, rather than by having to learn the exhaustive list of properties that define it. The study of 'natural' categories for Rosch may then begin with perception rather than logic, with typicality valued more than logical relations.[5] Therefore, the apprehension of a taxonomy begins with the elements exhibiting the highest degree of correlation with the world as given to perception, what she calls the *basic categories*.

This has significance to image classification. According to Rosch, we tend to choose basic categories *dog* and *cat* because they allow for a strong differentiation from their parent (i.e. *domestic animal*) and because the difference between *dog* and *cat* offers more immediate contrast than the difference between their subcategories (i.e. *difference between breeds of dogs or cats*). In addition, basic categories correspond to an economic optimum: they are the categories that require the least effort to reach.[6] To climb a taxonomic tree to reach a more abstract level or to go down to access a finer-grained class requires additional effort. In this respect all levels do not have the same 'inclusive' quality and they require extra work with lower information gain. We can see how the interest in the glance and a theory of classification rooted in basic categories may reinforce each other. The eye speaks to the brain in a structured fashion: there is an immediacy in early vision, a privileged access to the world. This can be located in a very precise interval and captured experimentally. Here cognitive psychology helps produce a model of vision where time is the intervening factor that can be submitted to experimental variations. What is perceived in that perceptual interval is a gist, a percept which relates to the first contours of the basic category objects. In early vision, the world opens up itself to the viewer at a mid-level perception. This moment at the threshold of consciousness, where perception is incipient, can be framed and measured temporally, transformed into visual stimuli and mapped onto a taxonomic scheme.

Photographic Pipelines

At this stage, we can sense already how such a subject is congruent with the viewer described by Li in the introduction to this chapter: a subject who has the capacity to classify two images per second, a subject psychologically and cognitively compatible with the scale and environment of annotation. The encounter between engineers and psychologists in the computer vision lab goes beyond an exchange of concepts. Cognitive psychology's method is experimental. Concepts come with measurements and technical devices, and experiments need actual resources. A crucial meeting point for computer vision scientists and psychologists is experimental craftwork. As visual stimuli do not exist in nature, they must be fabricated: to conduct experiments, images must be collected. Researchers such as Antonio Torralba, Pietro Perona and Li introduce a significant change in the provenance of images used in datasets for computer vision, and the elaboration of datasets subsequently impacts their approach to stimulus design. In a paper introducing the Caltech-256 dataset[7] assembled from photos found on the Internet, Perona and colleagues wrote that the dataset's quality was grounded in 'a diverse set of lighting conditions, poses, backgrounds, image sizes and camera systematics' (Griffin et al. 2007). This was contrasted with experimental psychology, where subjects were generally shown images after irrelevant information was discarded such as line drawings, or photographs of objects which had been shot by professional photographers against a neutral background to ensure a 'clean' descriptive image[8] (Fei-Fei et al. 2007). These options are deemed too different from the everyday visual experience of the subjects. They wanted to select 'real-world scenes' for their datasets as well as for their experiments, and the source for such scenes was the Internet.

By choosing images exhibiting such a diversity, the researchers hoped to avoid the introduction of representational bias. It is worth lingering on the meaning of the term bias in this context. The researchers involved in the study of scene understanding want to design a device that correlates specific micro-temporal variations in visual perception to mid-level classification. To design the proper stimulus, they need to acquire an imagery which corresponds to this level of description as accurately as possible. A line drawing would represent a different level of description. Lacking details and context, it would refer to higher levels of abstraction. The meaning of bias here does not so much concern a subjective interpretation of what is represented. A line drawing or a professional photo shoot are "biased" against a level of description. By acquiring images using search engines, they do not aim to eliminate bias, but to "average" it. To average bias does not mean to reconcile disparate subjective views on a same object. It means to select the images a majority of web users see as corresponding to basic categories such as bicycle, umbrella or flute player.

A quick visual example may explain how the search engine handles the difference between levels of description. Figure 4.1 shows the results of a search for "quadruped" in Google Images, while Figure 4.2 shows the results of a search for cat. The query "quadruped" displays images of people on all fours along with animals and even robots, whereas the query "cat" displays homogeneous imagery of the animal. But the difference is not only a matter of what is represented, there is also a striking difference in photographic registers. Quadrupeds display an isolated figure against a uniform background: they are didactic pictures, professional photo

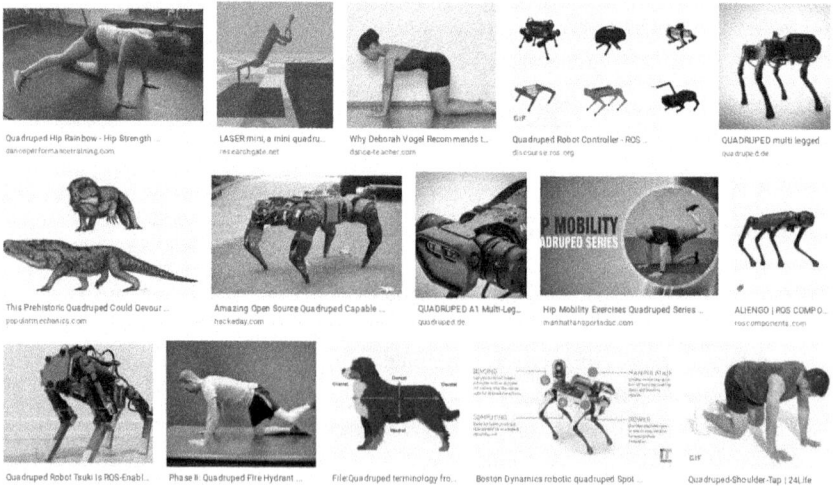

FIGURE 4.1 A screenshot of image search results corresponding to the query "quadrupeds"

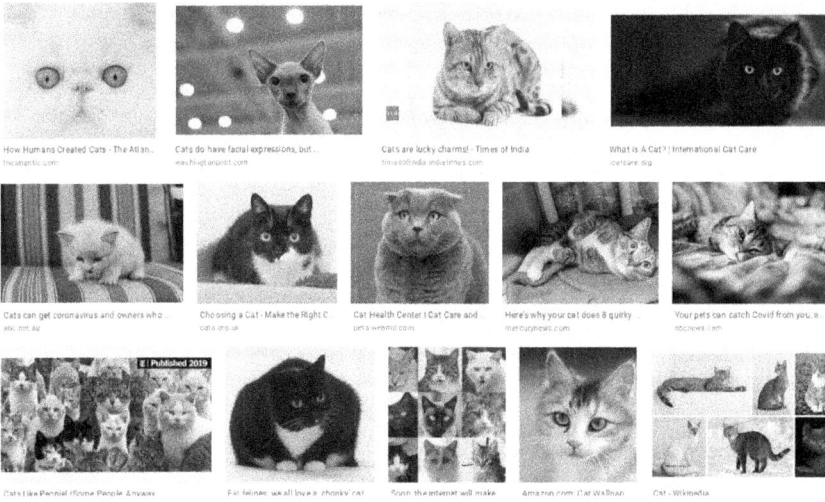

FIGURE 4.2 A screenshot of the image search results corresponding to the query "cats"

shoots or 3D reconstructions. In contrast, the results of the query "cat" show the animal in domestic settings, among familiar objects and in company of people or other animals. Here we have images valued for their "cuteness," exchanged in photo sharing platforms made by semi-professional photographers and amateurs. The researchers use the visual search engine to correlate basic category terms to candidate images, to find visual equivalents to specific taxonomic levels, but also to figure out which photographic register corresponds to a mid-level entry in a given classification scheme. Here a level of abstraction is not merely related to a distinct representation, but also to photographic apparatuses and modes of circulation. This means that the technical chain created by computer scientists to acquire the stimuli has more than a mere logistical function: it provides an epistemic function as well. The subject of vision is anchored to the world through mid-level perception and the world accessed through mid-level perception is the world seen through Flickr and Photobucket.

The Social Ontology of the Computer Vision Lab

The computer vision lab is productive: it produces a viewer whose perception follows a micro-temporal timeline, who catches the gist of a scene in a glance, and is able to map this percept to mid-level categories. These are not just abstract concepts: the computer vision lab enacts a world mobilising material entities and devices. Micro-temporal vision could never be correlated to mid-level categories without the selection of images pertaining to specific photographic registers. When computer scientists engage with cognitive psychology, they are ontologically active and logistically creative at the same time. To produce its concepts and validate them, cognitive psychology needs to design environments in which variables can be isolated, controlled and measured. The central task of the experimentalist is an inductive one: the identification of a factor that can be reliably linked to variations in other factors in the process of perception. This means to assign roles to experimental subjects and ensure proper control of their behaviour. An experimental device is not a purely hypothetical device, it needs to function and to function it needs to put subjects to work. In this context, to experiment is to manage subjects. Semiosis as studied in cognitive psychology needs to be mapped onto actors and devices, each with a specific task and an apparatus to measure their outputs.

As variability subtends the discovery of the causal mechanism, the agents that will embody the variables in the experimental settings must be cast with care. Unlike other experimental fields like "fundamental" or "hard" sciences, experimental psychology deals with human subjects, and creating a field of interactions with subjects means enabling a form of sociality. Sociality in the lab is enabled on a mode of severe reduction to enable experimental control, where the computer vision lab is composed of discrete individuals that can be treated as isolated atoms. The bare individuals taking part in an experiment are interchangeable, and their interaction is deemed irrelevant to the study of perception. As we have seen, these individuals are presented as ready to perceive, and the world is available to them. The world is made of concepts "imaged" through vernacular photography. Computer scientists as experimentalists have a predilection for imagery found on the internet as it represents "direct" access to common objects and beings unfettered by the conventions of professional photography (Fei-Fei et al. 2007). To a large extent, the recognition of the represented object happens as a transparent process, hence its photographic mediation can be ignored.

Subjects' perception is defined in atomistic terms too: perception can be broken down into discrete units, fixations, that can be quantified. In the computer vision lab, scientists study subjects who perform their tasks in isolation, and sociality is obtained through the mediation of an apparatus that collects the responses and processes them. The mode of consensus that can be obtained from the subjects is a process of averaging that dispenses with direct interaction. For instance, in the 2007 experiment discussed above, subjects describe photographic stimuli individually, not collectively, and they have no means to discuss their experience with other participants. In this respect the experiment does not offer any explicit feedback: subjects produce descriptions, but have no occasion to review their work or to engage in a discussion with the experimenters. Their knowledge is understood as basic and static: nobody expects them to improve, to learn from their past interactions, to get better at what they do. Experimental subjects are expected to mechanically respond to stimuli. In this way, the ontology adopts a reductive version of automatism limited to mechanical repetition. The course of an experiment is reduced to a concatenation of discrete sequences, and recognition presupposes that the work of cognition has already happened prior to perception. Recognition is understood as a way of matching what is given to the senses with already acquired categories. The privileged categories are Rosch's basic categories: the categories that correlate with the lowest perceptual effort and which form the subject's perceptual horizon. Finally, perceiving subjects are understood as "wired" perceptual agents, as cybernetician Stuart Russell remarks (Russell 1996, p. 111). The subject who identifies a piece of meat, a fork or a kitchen in a few milliseconds is culturally ("soft") or biologically ("hard") wired to the world, and their act of perception is validated through immediate reward rather than deliberation. The use of the term reward underscores the manner a correlation is produced: not by explicit negotiation and contractual specification, but by cueing and reinforcement.

I started my analysis of the computer vision lab as a site where scientists were studying human vision to make it tractable to algorithms. This led me to trace its conceptual background and its devices, its model of vision, its taxonomies and photographic registers, in order to finally to articulate its social ontology. At this stage, we better understand how the social ontology of the computer vision lab relies on an assemblage that goes beyond scientific concepts to include material devices and specific photographic registers. It is time to turn to the question that logically follows: what happens when what is tested in the lab is translated into the production environment? How do the ontology and the relevant parts of the experimental device travel together?

The Entanglement of the Lab and the Annotation Platform

Li and her colleagues' interest in cognitive psychology operates from two perspectives. On the one hand, they look for inspiration in the perceptual models the discipline provides. As machine vision expert David Lowe (Lowe 1999) writes, human vision is the proof that vision works. In this view, to start from existing models gives a foundation to invent artificial vision and to define what visual cognition consists in (for example, concepts such as the gist, the glance, mid-level perception and basic categories). On the other hand, busy with dataset creation, computer vision scientists look into cognitive psychology to understand how they can optimise the viewing and description of images so necessary to the annotation of their training sets. As I argue here, to apply what is discovered about vision in the lab to the environment of annotation requires a huge work of translation. However, to understand how the environment converges with the social ontology of the computer vision lab, a schematic description of the environment of annotation is first in order.

Machine vision would not exist in its current form without the mediation of different platforms providing free or cheap labour. Without them, it would not be able to mobilise the large workforce needed to annotate billions of images. Annotation is the cognitive labour that consists in bridging what computer scientists name the "semantic gap": a gap between the photograph as a visual surface that can be interpreted semantically, and the photograph as data. Bridging this gap is intimately related to the versatility of the networked image which operates simultaneously as a picture and as a collection of bytes.[9] To bridge this gap is a labour-intensive activity. Amazon Mechanical Turk (AMT) is one of the most popular examples of such environments (Irani 2015). AMT is an online platform acting as a broker between workers (called Turkers) and employers (called requesters). Workers apply for jobs called Human Intelligence Tasks, or HITs. These jobs are tasks deemed simple for a human, and complex for a computer. Turkers perform micro-tasks like selecting the colour of a garment or naming objects in a scene. In a Taylorist fashion, large work processes are decomposed in individual tasks that can be executed in parallel, and workers on the platform are anonymous and are treated as interchangeable units. This allows requesters to recruit a flexible workforce according to the needs of production. On AMT, advantage is clearly given to requesters: anonymity and lack of contact between the workers is the result of the platform's design to prevent the workers to getting to know each other and to unionise.

In the context of huge dataset projects such as ImageNet, Turkers interpret, filter, clean and verify thousands of images. They have to accomplish these tasks in the blink of an eye. If Turkers want to make a living or just a semi-decent income from their labour, they need to work at a pace that barely allows them to see the images. From this perspective, a speed of vision is built in the platform economically. For the annotators, structurally, the glance is the norm. Speed also corresponds to the evergrowing appetite from the software industry for training sets that can be delivered on demand; which means that legions of workers are mobilised intensely for a short period of time. Through the interface of the AMT, requesters are managing the cadence of the annotation work; they want to ensure the workers proceed fast enough to match production deadlines. The cost of the labelling effort leads computer vision researchers to approach visual content in the form of informational currency and attention scarcity. As the volume of requests augments, the unit of measurement for a labelling task moves towards the millisecond rather than the second.

With this description of the annotation environment in mind, let's revisit the social ontology proposed earlier and see how it resonates with an environment such as AMT. The first trait of this ontology is the understanding of the social as composed of isolated atoms. In the lab, the interactions between perceiving subjects are deemed irrelevant to the study of vision and the experimental device prevents them from controlling independent viewing subjects as isolated variables and measuring their behaviour. On the AMT platform, workers are isolated and their interaction is prevented for reasons of social control and productivity. Consensus therefore is treated as an averaging: an averaging of responses or an averaging of HITs, not as a negotiated process. Neither the lab nor the platform provides explicit feedback. *A model of separated individuals and consensus obtained through averaging is shared by the lab and the platform.*

Secondly, cognitive psychology – a laboratory science – translates concepts into material experimental situations. In these situations, subjects are assigned roles which are valued differently, reflecting the part of the model of vision they represent. Here, computer scientists as experimentalists consider early vision as mechanical and consequently, they treat their experimental subjects as merely responding to stimuli. When computer scientists translate their experimental layout to the annotation environment, they treat the workers accomplishing early vision tasks as automata who perform a low value work. Ranges of variability and grades of effort measured in the lab are translated in worker's management techniques and levels of remuneration. In this model early vision means low wages. Because "late" vision with its higher level of interpretation is considered a skill requiring experience, it is considered too expensive for production. *The model of vision produced by the computer vision lab is translated into an economic model for the annotation platform with corresponding hierarchies.*

Thirdly, both subjects and workers are ready to perceive a world available to them through photographs. To perceive in the lab and on the platform means to correlate visual stimuli sourced from vernacular photography to basic categories, and they are expected to perform acts of recognition understood as immediate. *To recognise* is understood in this context as the mechanical application of already acquired knowledge onto incoming visual data. There is an assumption that photographs obtained through search queries can be treated as transparent containers of visual nouns and these visual nouns correspond to mid-level entries in taxonomies. Because photographs can be treated as transparent and because basic categories correspond to acts of classification requiring the least effort, *the work performed in the lab and on the platform can be effectuated at speed.*

But in order to treat the photograph as a transparent medium, engineers engage with its ontology as well as its materiality. A photograph found on Flickr by a search query doesn't become a dataset item by magic. To be included in the dataset, a digital file is excised from the Flickr environment. In many datasets such as ImageNet and COCO, the comments do not travel with the file, neither do the albums, metadata information, or the author's name. The Flickr photo, previously tagged by the author or the community, enters the lab anonymously. It is now stabilised and positioned as visual data "imaging" a category entry. The object once valued for its pictorial or affective quality is now understood as a taxonomic container. This operation is far from innocuous. The photograph is enacted differently: it is realigned, and this realignment requires a considerable amount of work. Computer vision researchers understand very well that vision can be many things and that, to be studied, it needs to be held in place. Photography is what gives seeing its definition: to see is to see a photograph within a very specific microtemporal range. To enable vision at speed, the photograph must comply with the expectations of the viewer, in which a real-world scene might be grasped in an amateur snapshot. *The lab and the annotation environment share an understanding of the vernacular as stimulus.*

The computer vision lab offers more than a social ontology to scientists who design environments of annotation. It also offers an apparatus: the experimental device, which in turn operationalises the social ontology. As we have seen, the lab provides an environment where perception can be addressed in terms of variability: it offers the means to control the variables deemed relevant by the experimenters and the means to hide other sources of change which it treats as unwelcome interferences. In this respect, the experimental apparatus provides a material environment already parametrised to produce viewers at speed, not just to study them.

Translation

It is important here to notice that the translation from the lab to the platform is not a smooth process. It requires ingenuity. The common ontology permeating both environments offers a ground to computer vision scientists, but its translation remains adventurous. Crucially, the translation is not unidirectional. As Sarah Kember (2014, p. 194) has noted there is a circularity inherent to computer science and computer vision in particular. In the traffic between computer science and the psychology of vision, what begins as an inspiration to design algorithms ends up as a technique to optimise the humans who train them. With each new cycle of investment and innovation, the enfolding of the computer vision lab and the annotation platform increases. For instance, as is the case for numerous academic disciplines, the separation between the institutions of education and research (e.g. lab) and the environments of production (e.g. the software industry) are less and less separated. When computer vision research became increasingly dependent upon industrial sponsorship and aligned to its agenda, key academic figures of the were are hired to important positions in high-profile companies (i.e. Fei-Fei Li at Google,[10] Yann LeCun at Facebook[11] or Andrew Ng at Baidu[12]). Over a few years, the landscape of research in AI and computer vision has been remodelled entirely and distributed over a complex network of sites of different natures. AI is the object of sector deals and strategic plans in which sources of investment have diversified. Successful computer vision scientists are at the same time directors of research centres in universities, project leads in major software companies and policy consultants for states and think tanks. The traditional separation between the lab and the production environment does not hold anymore, and neither does the separation between the academy and industry. In this respect, an environment such as the AMT represents a "factory without walls" as much as a lab without borders.

In the context of AI industries, the evolution of experiments led by Li and colleagues are symptomatic. Early experiments were meant to study human vision in terms that made it algorithmically tractable. Confronted with the rhythms of the annotation environment and the need to label images at speed, Li applies the experimental techniques of the lab to the management of workers, and the optimisation of Turkers' annotations. Li's experimental work provided an ontology compatible with the scale of the environment of production. Now that the lab blends into the environment of production, the motivation to develop new experiments is no longer informed by the study of human vision to inspire the design of algorithms. It is instead the optimisation of dataset production because labelling doesn't scale up as quickly as the volume of data to be annotated. What motivates a series of new experiments is the difficulty for the annotation process to match the production of industry-sized datasets. In a recent experiment outlined in "Embracing Error to Enable Rapid Crowdsourcing" (Krishna et al. 2016), Li studies how workers submitted to RSVP techniques can augment their productivity if a higher rate of errors is tolerated. Workers are shown a flow of images displayed on a screen for a few milliseconds. They are asked to press a key when a given object appears. As the flow is uninterrupted, for example when the worker presses a key, a number of sub-sequent images have already appeared on the screen, leaving the worker unable to second guess or correct their choice. By tolerating errors and correcting them post hoc, the authors of the paper conclude their approach is capable of achieving an increase of speed in the order of ten times the magnitude (Krishna et al. 2016, p. 1).

The same RSVP technique originally used in the lab with psychology students to further scene understanding research is now mobilised on Amazon's platform to augment the productivity of annotators. In this respect, research and production blend into each other: experimental techniques have moved to the industrial environment, and workers replace graduate students. Experiments are conducted in place and Turkers oscillate between their status as workers and subjects. Workers become experimental subjects in their own work environment. In this way, Li's 2007 Caltech experiment testifies to the difficulty of translating the experimental model in the annotation environment, in which the common ontology reveals its cracks under the pressure of production. The AMT worker cannot simply be reduced to a mere naive viewer as Li's grad student had been. They don't just need to graduate: they work to be able to eat, and have a roof under which they can sleep. For the worker errors matter as they transform their percepts into rewards, and recognition is not a discrete act that can be repeated mechanically. Workers recognise things in the images they are administered, but they are aware that they are also monitored. As one's ratings on the platform evolves, one is also able to apply for better paid jobs. Workers also become able to improve their position and sub-rent their account to others, turning anonymity and interchangeability to their advantage (Schmidt 2020). As the worker is recognised as a good Turker, they may become an intermediary for other workers. Naiveté changes camp. If the error rate associated with their account increases, the workers risks losing their status, access to better jobs and the ability to sub-rent. Experimental subjects become workers, workers become experimental subjects yet workers also become entrepreneurs capitalising on their own reputation to outsource the task further down.

To increase speed of annotation at the risk of provoking errors, the cooperation of the worker is required. On the AMT platform it is not enough to ask them to go faster: the worker needs to behave more as an experimental subject detached from the consequences of their response. For computer scientists, the solution to this new development is not easy, and it is necessary to design an apparatus that primes workers to refine the cadence of annotation. The worker's anticipation of their employer's judgement is handled in the same terms, and becomes a factor for which a degree of variability needs to be figured out experimentally. This triggers a redesign of the environment where errors are now "embraced" and are integral part of what is expected from the worker as a sign of their productivity.

Having discussed the lab's enfolding with the annotation environment, we now reach the end of our journey. Through this process a subject has been constructed who will verify 14 million images in a glance, relieving Li's psychology students of their role in the Caltech experiment, allowing them to graduate. This chapter has positioned the computer vision lab as the site where this subject is engineered.

The existence of such a subject requires a model of vision defined by a specific temporality, the glance. This experimental subject is bound to a theory of classification that privileges a level of abstraction that requires the least cognitive effort: the basic categories. As perception is modelled on the paradigm of early vision, visual cognition in the computer vision lab is deemed effortless. Subjects have immediate access to visual nouns embedded in images. The subject of vision displays a natural ability which does not qualify as knowledge proper, while the model of perception also requires a selection of photographs congruent with a level of abstraction: the vernacular photograph. In the computer vision lab, where machine vision scientists meet cognitive psychologists, they enact a social ontology and compose a material device that help manipulate subjects and objects experimentally. As we have seen at the end of our journey, these subjects and objects do not stay in the lab. The relation between the industrial environment of annotation of machine vision and the experimental practice of computer scientists is epistemic and managerial: the experimental device acts as a proxy between epistemic production in the computer vision lab and cognitive extraction in the environment of annotation.

Conclusion

In analysing the computer vision lab, I have established that there is a factory in the experiment, and there is also an epistemic device at work in the annotation environment. What is at stake in the experiment is more than the observation of the perceptual properties of human subjects. As the experiment is mobilised to meet production imperatives and align subjects to a scale, it provides a management template for the annotation environment of computer vision. Because the experimental device positioned subjects as merely responding to "stimuli," workers could then be asked to verify photographs in a glance for a few cents. In this sense, the computer vision lab articulates and operationalises a social ontology that imbues the social organisation of the annotation platform. This ontology however does not stand on its own: the encounter between machine vision researchers and cognitive psychology produces material configurations and techniques that enable the translation of concepts and subjects in a production environment. Such a translation is by no means mechanical, and involves a redefinition of the cognitive objects from the lab in the operational objects of production.

By analysing the increasing entanglement between the lab and the platform within a cycle of investment and innovation, my aim is to emphasise the importance of approaching machine vision as more than mathematics and code. And equally, to emphasise the importance of treating the alliance of computer vision with other scientific disciplines as more than an exchange of ideas. The computer vision lab is a site where new epistemic agents are introduced during a period when annotation labour is outsourced and photo sharing platforms are defining what counts as stimulus. In the computer vision lab, forms of cognition and labour are used and made invisible at the same time. The epistemic configuration of machine vision is distributed over cycles of image capture, annotation, aggregation in datasets, experimentation, classification, measurement, investment, traversing the lab and the production environment. This approach is crucial because it changes our understanding of the sites where knowledge is produced, and where imagining critical interventions is possible. If we consider machine vision as a distributed process, every juncture, every connection in the process is an entry point for enquiry and for interventions that may affect its programmability. The question of what counts as knowledge cannot be separated from the world in which machine vision is enacted and that it helps enact. To alter the course of computer vision's development therefore means more than coding a different algorithm or selecting better photos in a dataset: it requires an intervention into the social ontologies of machine vision, in its division of labour, and the relations between its research institutions and the industrial environment of its production.

Notes

1 In this chapter, the terms computer vision and machine vision are used interchangeably.
2 Main collaborators of Li relevant to this text are Pietro Perona, Christof Koch, Jia Deng and Antonio Torralba.
3 Key datasets in computer vision are currently composed nearly exclusively of photographs. ImageNet or Caltech 265 actively prohibited other forms of representation.
4 Rosch's work on categories is presented as a milestone in publications as diverse as cognitive psychology syllabi like Braisby and Gellatly's *Cognitive Psychology* (2012) and STS literature such as Bowker and Star's *Sorting Things Out* (2000).
5 "The perceived world comes as structured information rather than as arbitrary or unpredictable attributes. Thus maximum information with least cognitive effort is achieved if categories map the perceived world structure as closely as possible" (Rosch 1978, 2).
6 "The task of category systems is to provide maximum information with the least cognitive effort" (Rosch 1978, 4).
7 The Caltech-256 Object Category dataset is a collection of 30,607 images culled from Google images and organised in 256 categories by Pietro Perona with Gregory Griffin and Alex Holub in 2007.
8 All the scare quotes in this paragraph indicate expressions used by the researchers in the report.
9 For an in-depth conceptualisation of the image at the interface of representation and code, see "Affordances of the Networked Image" (Cox et al. 2021).
10 Li was chief scientist of AI/ML at Google Cloud and has been appointed to Twitter's board as independent director. See www.cnbc.com/2020/05/12/twitter-adds-former-google-vp-and-ai-guru-fei-fei-li-to-board.html.
11 Le Cun is chief scientist at Facebook https://qz.com/1186806/yann-lecun-is-stepping-down-as-facebooks-head-of-ai-research/.
12 Andrew Ng worked until 2017 as AI chief at Baidu, see https://medium.com/@andrewng/opening-a-new-chapter-of-my-work-in-ai-c6a4d1595d7b.

Bibliography

Arnheim, Rudolf. 1969. *Visual Thinking/Rudolf Arnheim*. Berkeley, CA: University of California Press.

Boden, Margaret A., ed. 1996. *Artificial Intelligence*. 2nd ed. San Diego, CA: Academic Press.

Bowker, Geoffery C., and Susan Leigh Star. 2000. *Sorting Things Out: Classification and Its Consequences*. Cambridge, MA: MIT Press.

Braisby, Nick, and Angus Gellatly, eds. 2012. *Cognitive Psychology*. Oxford: Oxford University Press.

Cox, Geoff, Annet Dekker, Andrew Dewdney, and Katrina Sluis. 2021. "Affordances of the Networked Image." *The Nordic Journal of Aesthetics* 30 (61–62): 40–45. https://tidsskrift.dk/nja/article/view/127857.

Deng, Jia, Wei Dong, Richard Socher, Li-Jia Li, Kai Li, and Li Fei-Fei. 2009. "ImageNet: A Large-Scale Hierarchical Image Database." In *2009 IEEE Conference on Computer Vision and Pattern Recognition*, 248–55. https://doi.org/10.1109/CVPR.2009.5206848.

Fabre-Thorpe, Michele, Arnaud Delorme, Catherine Marlot, and Simon Thorpe. 2001. "A Limit to the Speed of Processing in Ultra-Rapid Visual Categorization of Novel Natural Scenes." *Journal of Cognitive Neuroscience* 13: 171–80.

Fei-Fei, Li, Asha Iyer, Christof Koch, and Pietro Perona. 2007. "What Do We Perceive in a Glance of a Real-World Scene?" *Journal of Vision* 7 (1): 10. https://doi.org/10.1167/7.1.10.

Fei-Fei, Li, Rufin VanRullen, Christof Koch, and Pietro Perona. 2002. "Rapid Natural Scene Categorization in the Near Absence of Attention." *Proceedings of the National Academy of Sciences of the United States of America* 99 (14): 9596–9601. https://doi.org/10.1073/pnas.092277599.

Fei-Fei Li. 2011. "Large-Scale Image Classification: ImageNet and ObjectBank." YouTube Video, 1:00:46, 18 May 2011. www.youtube.com/watch?v=qdDHp29QVdw.

Griffin, Gregory, Alex Holub, and Pietro Perona. 2007. *Caltech-256 Object Category Dataset*. California Institute of Technology. https://resolver.caltech.edu/CaltechAUTHORS:CNS-TR-2007-001.

Hayles, N. Katherine. 2020. *Unthought: The Power of the Cognitive Nonconscious*. Chicago, IL: University of Chicago Press. https://doi.org/doi:10.7208/9780226447919.

Irani, Lilly C. 2015. "The Cultural Work of Microwork." *New Media & Society* 17 (5): 720–39. https://doi.org/http://dx.doi.org/10.1177/1461444813511926.

Kember, Sarah. 2014. "Face Recognition and the Emergence of Smart Photography." *Journal of Visual Culture* 13 (2): 182–99. https://doi.org/10.1177/1470412914541767.

Krishna, Ranjay, Kenji Hata, Stephanie Chen, Joshua Kravitz, David A. Shamma, Fei-Fei Li, and Michael S. Bernstein. 2016. "Embracing Error to Enable Rapid Crowdsourcing." In *Proceedings of the 2016 CHI Conference on Human Factors in Computing Systems*, 3167–79.

Lettvin, Jerome Y., Humberto Maturana Romesin, Warren S. McCulloch, and Walter H. Pitts. 1959. "What the Frog's Eye Tells the Frog's Brain." In *Proceedings of the IRE* 47 (11): 1940–51. https://doi.org/10.1109/JRPROC.1959.287207.

Leys, Ruth. 2011. "The Turn to Affect: A Critique." *Critical Inquiry* 37 (3): 434–72. https://doi.org/10.1086/659353.

Lowe, David G. 1999. "Object Recognition from Local Scale-Invariant Features." In *Proceedings of the International Conference on Computer Vision-Volume 2*, 1150. ICCV '99. Washington, DC: IEEE Computer Society. http://dl.acm.org/citation.cfm?id=850924.851523.

Marr, David. 1982. *Vision. A Computational Investigation into the Human Representation and Processing of Visual Information*. Cambridge, MA: MIT Press.

Massumi, Brian. 2002. *Parables for the Virtual: Movement, Affect, Sensation*. Durham: Duke University Press.

Perona, Pietro. 2010. "Vision of a Visipedia." *Proceedings of the IEEE* 98 (8): 1526–34. https://doi.org/10.1109/JPROC.2010.2049621.

Pylyshyn, Zenon W. 1999. "Is Vision Continuous with Cognition? The Case for Cognitive Impenetrability of Visual Perception." *Behavioral and Brain Sciences* 22 (3): 341–65.

Read, Jenny C. A. 2015. "The Place of Human Psychophysics in Modern Neuroscience." *Neuroscience* 296: 116–29. https://doi.org/https://doi.org/10.1016/j.neuroscience.2014.05.036.

Rosch, Eleanor. 1978. "Principles of Categorization." In *Cognition and Categorization*, edited by Eleanor Rosch and Barbara Lloyd, 27–48. Hillsdale, NJ: Erlbaum.

Russell, Stuart, 1996. "Machine Learning." In *Artificial Intelligence*, edited by Margaret Boden, 89–133. San Diego, CA: Academic Press.

Schmidt, Florian. 2020. "Unevenly Distributed." *Unthinking Photography: The Photographers' Gallery*. https://unthinking.photography/articles/unevenly-distributed.

5

WAYS OF MACHINE SEEING AS A PROBLEM OF INVISUAL LITERACY

Geoff Cox

"Seeing comes before words." John Berger's *Ways of Seeing* opens with this statement (1972, p. 7). Although words are used to make the case, the meaning is made clear that "the reciprocal nature of vision is more fundamental than that of spoken dialogue" (1972, p. 9).[1] By this, Berger explains how seeing reinforces an ocular-centric paradigm of knowledge and power founded in Western modernity and colonialism, based on the relative positions of who is looking and being looked at. That every image embodies a way of seeing, as he puts it, demonstrates how, even without words, images support an elite worldview, and hence the necessity of visual literacy to draw attention to the underlying ideologies at work (including, for instance, in *Ways of Seeing*, representations of the objectified female body in art history and commercial advertising, or indeed how class privilege is reinforced in portraiture). To stress the point, Berger's closing remarks to the first episode of the television series reflects on the medium through which his ideas were made public, and what is missing at the level of socio-technical infrastructure:

> But remember that I am controlling and using for my own purposes the means of reproduction needed for these programmes. The images may be like words but there is no dialogue yet. You cannot reply to me. For that to become possible in the modern medium of communication access to television must be extended beyond its present narrow limits. Meanwhile, with this programme, as with all programmes, you receive images and meanings which are arranged. I hope you will consider what I arrange but be sceptical of it.[2]

The critical analysis of visual culture, to which Berger's essay continues to be a key reference, highlights the underlying conditions that allow us to see how visuality is constructed, and by extension how knowledge of the world is produced – and consolidated into worldviews. Seeing is an effective way in which power differentials

DOI: 10.4324/9781003095019-8

FIGURE 5.1 John Berger's *Ways of Seeing* seen through image recognition software

are legitimatised, and yet, as Berger points out, "the relation between what we see and what we know is never settled" (1972, p. 7). Yet this broader understanding, or literacy, has become more and more difficult to implement as networked images no longer simply represent things in the world but are an active part of invisible visual culture, and as such exhibit new forms of distributed power.

This essay sets out to explore how the relation, identified by Berger, between what we see and what we know, has been further unsettled.[3] How has visual literacy been transformed by developments in computer vision, underpinned as it is by developments in machine learning? It is organised into two main parts. First, it asks what is at stake in the analysis of culture through literacy, and subsequently in current debates around computational literacy. In the second part, an expanded notion of visual literacy is problematised by developments in computer vision, and what is referred to as *machine ways of seeing*. Several questions arise consequently. When most images are made by machines for other machines, and distributed across planetary networks, and part of vast annotated datasets, how are worldviews reinforced differently, and what kind of literacy applies, if at all? To what extent are the relations between words and images transformed by computational processes, and what are the implications for the theoretical frameworks we apply?

The uneven relation between words and images are reinforced by the book cover image (Figure 5.1), which not only repeats the opening sentences of its first page but also tellingly reproduces René Magritte's painting *The Interpretation of Dreams* (1927), in which images and their descriptive labels contradict their representational capacities. For this essay, the one you are reading 50 years later, the analogy to object recognition in computer vision is made explicit in Trevor Paglen's *The Treachery of Object Recognition* (2019) – homage to René Magritte's *The Treachery of Images* (1929) – in which the original exhibition poster is overlaid with green rectangles and classification labels.[4] The software *sees* that this is a "red and green apple." How and what computers recognise in an image, and indeed what they misrecognise, neatly demonstrates the difficulty that underscores any seeing event and how we might conceive of literacy differently in the context of non-human languages and machine intelligence, even if the underlying ideology remains relatively unchanged.

For Berger, it was the ambiguity of the image that generated its power of signification, understood at the time of writing, in the early 1970s, through the methodologies of semiotics and structuralism, as well as the historical materialism of Walter Benjamin (whose "artwork essay" was translated into English around the same time, and who Berger credits for the inspiration of the first episode of the TV programmes).[5] Moreover, the notion of visual literacy is a paradoxical notion, as if seeing can be equated with reading, reinforced as it has been by the linguistic metaphor at the heart of much critical theory. If both signs and images have taken on an even more prominent role of contemporary culture – the effect of the coming together of semiotics and capitalism (aka *semiocapitalism*), and in which there are ever more points of view, perspectives, attention, visual metaphors, screens, infrastructures, and so on – might we regard these as ever proliferating examples of "visual hegemony," in which the complex operations of algorithmic and data processes that are based on existing prejudices are even more hidden?[6] The influence of the concept of hegemony in Berger's work is apparent in his analysis of works of art, but its applicability to computer vision remains in question,[7] although clearly it can help to highlight some of the vested interests.

More to the point, the relation between what is visible and the names that we give to what we see to make sense of it persist as a problem of literacy, rooted in the tendency to conflate representations with the things that they represent (whereas we know the relation to be arbitrary if we follow the lessons of semiotics). But to what extent does media literacy remain up to the task of analysing contemporary visual culture and to understand the ways in which signs have been incorporated into the operations of capitalism? In addition, what are the misrecognitions afforded by theoretical concepts that seem to assume images to be singular or human-centred or indeed like language? This is clearly an issue when images are distributed across networks, and when they are made by machines for other machines, and at enormous scales. Moreover, images exhibit the ability to act in the world, and upon us, famously articulated by Harun Farocki's notion of "operational images" (2004). They no longer simply represent things in the world but are an active part of invisible visual culture—part of an operation, as Farocki puts it, externalising new forms of distributed power that derive from the *eye/machine*. In the case of computer vision systems, they make judgements and decisions, and as such exercise power to shape the world in their own terms, which, in turn, upholds the argument that they embody new ways of seeing.

Seeing, then, can no longer be thought of as singular or indexical truth or reality, as it was mistakenly thought to be, but is indicative of a wider need to manifest authority and power, through distributed forms. But to what extent is this a question of literacy? Indeed, does literacy remain a useful descriptor for reading computational images, especially given its enthusiastic uptake in education policy and the creative industries, and its adaptation to other technological forms and cultural values (e.g., digital literacy or carbon literacy) – to the point where almost anything requires training to "level-up" social and economic status. Such questions over changing literacies, or whether literacy is a metaphor fit for purpose in contemporary culture, occupy the rest of this essay. But first, I briefly explain what is meant by literacy.

Uses of Literacy

As suggested, literacy indicates not only the cultural ability to read and write but more broadly demonstrates competence or knowledge of practices that allow users to maintain and build social imaginaries (Celiński 2019, p. 467). The technology of writing is an example of the ways in which human cultures have been transformed, from the first written scripts developed by the Sumerians (circa 3500 BC) to the printing press of the Middle Ages in Europe to artificial intelligence today. However, it should be remembered that the skill to read and write were once specialist skills, and not something that extended to the whole of society. The Latin roots *literatus/litteratus* refer to someone who is educated and "who knows the letters," but who also defines the letters, and more to the point, the political structures in their learning and application. To know and define letters clearly involves divergent competences, which can be non-written, algorithmic, and performative, but no less grammatical or arranged (to echo Berger).

To clarify further what is meant by literacy, I refer to Richard Hoggart's *Uses of Literacy* (published in 1957) that challenged many of the assumptions of what constituted *culture*, as previously the preserve of an elite (2009).[8] The subtitle of Hoggart's book confirms the class politics around forms of literacy and non-standard forms of expression, and how hegemonic forms of privilege – related to class, but also by extension, gender and race – are underscored by the ability to read and write in ways that affirm social status and cultural value. For instance, this is evident in human speech, in accents and dialects, choice of language, and so on, as well as in everyday voice-controlled devices such as Apple's Siri or Amazon's Echo that have preferential modes of address that similarly reflect class, gender, and racial stratifications.[9] The authority expressed in literacy is founded on the privileging of certain forms of language over others – who speaks, to whom, in what way, and under what conditions?

The politics of literacy was further developed in the interdisciplinary field of cultural studies (indeed the Birmingham School was founded by Hoggart), setting out to understand how meanings are produced within wider systems of power and control, even in the most everyday circumstances, and in recognition of the global expansion and significance of culture into all aspects of life. In this sense, literacy, although historically rooted in literary studies, operates as a broader metaphor for the active role of users in "encoding/decoding" media messages (Hall 1980). The politics of this expanded media literacy is that users can produce meanings through collective action (or counter-hegemony) rather than simply receiving the message as intended as if an uninterrupted flow of information. In other words, literacy is socially constituted, and takes material forms – and this is the case in visual culture and popular media, as well as computational information based as it is on material infrastructures and language.

In summary, literacy is a combination of individual skills, a material system, and a social practice – "useful" in terms of its wider application, which is enabled by its particular infrastructure, and constantly changing and being transformed by the development of inscription technologies. In her 2017 book *Coding Literacy*, Annette Vee reinforces the point: "Literacy is a widely held, socially useful and valued set of practices with infrastructural communication technologies" (2017, p. 27). If, as broadly agreed, there is a moral imperative to encourage everyone to learn to read and write, then surely to code too. Moreover, where to draw the limits and decide what constitutes reading, writing, or coding/programming, let alone the ability to do any of these competently (and demonstrate "good" literacy in any one of them)? What is at stake is an expanded understanding of literacy – the ability to read, write, *and program* – an enhanced understanding of the relationship between what words mean and do.

The arguments for coding literacy seem compelling but also raise concerns about the motivations around the kinds of literacy being introduced, and to what extent this deviates from some of earlier discussions around counter-hegemony. Here various initiatives such as online tutorials and websites come to mind, such as Codecademy.org and Code.org (ominously backed by Facebook's Mark Zuckerberg and Microsoft founder Bill Gates) and other educational platforms for the common good, particularly in the United States. MIT's Scratch is a good example of software designed to teach children how to code, or more to the point, the principles of "computational thinking" (the process of breaking down a problem into simple enough steps that even a computer would understand) (Wing 2006, p. 33–35). In this case, to be literate is to *think* mathematically or algorithmically, like a machine – and in the context of this essay, to see like one too. Whether this is useful remains in question, as clearly the development of literacy as a project can be seen to follow what Stefano Harney and Fred Moten (2013) have described as the proliferation of capitalist logistics through the management of pedagogy.

Vee's book challenges the assumption that literacies are necessarily assumed to be for the social good (2017, p. 2). More to the point, what literacy is, and what kinds of skills are required to become literate, have tended to be kept vague to be easily shaped for vested interests. In contrast, numerous examples are to be found of *illiterate* or non-standard forms of expression, such as the use of slang and creoles in speech and writing, but also *esoteric* programming languages that offer other forms of legibility. The illegibility is precisely the point to disrupt hierarchies – an example of which is "brainfuck,"[10] a Turing-complete programming language that only uses non-alphabetical characters ><+-.,[] as commands to confound human-readability. Through such examples, it becomes clear that writing is a form of action, and not simply a referent of thinking.

Indeed, writing is programming and programming is writing that instructs a computer about what to do in human-readable language, in turn translated by a compiler into something the computer can parse. In this way a simple human-readable instruction like "print, loop, end" can give a clear indication of what is taken place even if in fact there are multiple processes in operation. The phrase "literate programming" introduced by computer programmer Donald Knuth in 1984, indicates how a computer program contains an explanation of its logic in a natural language, such as English, interspersed with traditional source code, from which compilable source code can be generated. This is not only functional but also holds aesthetic potential for Knuth: "Literature of the program genre is performable by machines, but that is not its main purpose. The computer programs that are truly beautiful, useful and profitable must be readable by people" (1984, p. 97–111). This extends beyond the straightforward use of comments, or the naming of objects and abstractions to how they produce wider meaning. The ambiguity of the word "class" is an example: describing objects in programming as well as social stratification. Literacy can thus be seen to be enhanced both by computational forms and as a potential form of (class) action that helps to understand algorithmic or rule-based systems. An extract from Harwood's codework *Class Library* (2008) illustrates the point:

```
# We are left with no option but to construct code that
# concretizes its opposition to this meager lifestyle.
package DON'T::CARE;
use strict; use warnings;
sub aspire {
my $class = POOR;
my $requested type = GET_RICHER;
my $aspiration = "$requested_type.pm";
my $class = "POOR::$requested_type";
require $aspiration;
return $class- >new(@_);
}
1;
```

The detail here is important as programming is a very particular kind of writing, both a description of an action and the action itself (it says what it does). As such it cannot be divorced from the social and material conditions in which it is produced and distributed. It is not simply a new way of reading and writing but also a new way of thinking and understanding other codes. Literacy in this sense not only benefits those who acquire certain skills but also has potential wider cultural and social ramifications, helping to force coding out of its specialisation in certain disciplines and open its critical potential more widely. This should clearly not simply be the preserve of computer science, nor an elite group of specialist programmers who control the borders of media and networks. This is especially important when considering what new kinds of literacy are required to engage with contemporary inscription practices – not only text-based forms such as electronic writing and computer programming but also visual forms such as video conferencing and computer vision – as well as the broader infrastructures in which they operate. Berger's earlier comments on the medium of television comes to mind, and how new ways of seeing require new forms of literacy to account for the invisible realm of algorithms and database (infra)structures of computer vision. Every image embodies ways of seeing, but what ways, and how are they arranged, by whom and to what purpose?

Invisualities

Drawing on Berger, Nicholas Mirzoeff situates the intensification of the visual in culture is a symptom of the wider need to manifest authority through visuality (2011). What he refers to as the "right to look" is denied for some and not for others, and any act of seeing is met by a willingness to be seen, in other words is founded on reciprocity. He offers the historical example of "reckless eyeballing," the act of looking at a white person by a slave, looking at a figure of authority in a manner considered to be violent (Mirzoeff 2011, p. 482). We might immediately identify a parallel in the "broken metaphor" of "master" and "slave" in programming where one process exerts control over another process within a dependent relationship (Eglash 2007).[11] At its extreme, seeing becomes a "necropolitical" issue – to adopt the phrase by Achille Mbembé – posing the question of who lives and who dies (Mirzoeff 2011, p. 487).

The material destruction of human bodies and populations are increasingly performed from above, using drones for instance, as part of the "post-panoptic imaginary" that also separates the enemy as if in a computer game (Mirzoeff 2011, p. 488–489). Here we are further reminded that images can kill, as in the case of the POV (point of view) perspective of the bomb plunging towards its target – what Farocki calls a "suicidal camera" (2004). These "vision machines," as Paul Virilio put it, exemplify the paradoxical notion of an act of "sightless vision" (1994, p. 59) – "eye machines" that do not "see" as such, but instead they "read" the world according only to the logic of the model of world they know (based on the particularities of a specific image dataset for instance). It is a lopsided model of the world, a colonial worldview that correlates with other forms of cultural imperialism (such as the hegemonic dominance of English language) and one in which new forms of literacy are required that are not merely based upon representational paradigms. Image datasets confirm the problem in computer vision, wherein an algorithm constructs a worldview based upon its limited resources. An example is ImageNet, a large visual dataset used for visual object recognition, derived largely from amateur photography in North America and the annotations of precarious workers on

Amazon Mechanical Turk (Malevé 2019). In other words, in describing machines as able to *see* we adopt a shorthand for calculative practices that only approximate likely outcomes by using probabilistic algorithms and models that have already been built upon inherent human prejudices related to class, gender, and race (Crawford & Paglen 2019). When computer vision systems *see*, they exercise power to shape the world in their own image, which, in turn, is built upon embedded biases and the problem of generalisation. The relation between seeing and knowing is likely to be set at the lowest common denominator.

Reading source code to understand how a machine sees is not particularly revealing in itself – despite what was argued earlier – but rather requires a more networked and relational literacy. To reiterate the point, it is not simply a case of how humans see the world, or how they use machines to see (as in the case of photography), but how machines see and produce the world in their own terms. Machinic literacy is required to understand the implications of this more fully, how our worldviews are being reproduced, and how seeing is enmeshed with the "machinic unconscious"[12] – for not only what is seen but also for what remains unseen yet still operative. The principles of visual literacy similarly would thereby stress how seeing is not simply a way to perceive the world but also a way to act within it. But what about computer vision and its distinctive way of seeing, and how to gain access to its underlying ideological structures and effects?

The relation between what we see and what we know remains at the centre of this, echoing Berger's first words at the beginning of this essay in which he establishes the primordiality of the image. To repeat, "Seeing comes before words. The child looks and recognizes before it can speak. [...] The reciprocal nature of vision is more fundamental than that of spoken dialogue" (1972, p. 7 and 9).[13] This echoes the phenomenology of Merleau-Ponty, who also sets out the difficulties and contradictions of seeing: "It is at the same time true that the world is what we see and that, nonetheless, we must learn to see it" (1968, p. 4). But how do we learn to see, and more to the point how do machine learn to see? Clearly, they are not one and the same, and yet it is common to draw analogies between machine intelligence and cognitive development in humans, especially in children, following a (broadly constructivist) idea of learning as something informed by experience. Yet this can also appear superficial when applied to teaching a machine to see, as for instance, in the following example – cited by artist-researcher Nicolas Malevé – of Fei-Fei Li describing her insight into the development of ImageNet:

> If you consider a child's eyes as a pair of biological cameras, they take one picture about every two hundred milliseconds, the average time an eye movement is made. So by age three, a child would have hundreds of millions of pictures of the real world. That's a lot of training examples. So instead of focusing on solely better and better algorithms, my insight was to give the algorithms the kind of training data that a child was given by experiences, in both quantity and quality.
>
> *Fei-Fei cited in Malevé 2019*

The example presents a reductive equivalence between human and machine vision. Yet Malevé's concern is more about what is implied about training and learning in general. He points out that we are all involved in the process of teaching machines to look at images in everyday situations and describes the enormous amounts of training that takes place when we use everyday devices such as smart phones and computers. Yet his interest is not so much our complicity in these processes, but to investigate which pedagogical methods might be useful. What can we learn about learning from the dynamics of machine learning? In his words, how to "transform it and be transformed by it? Or, to formulate this in terms even closer to Fei-Fei Li's, how can we think productively about the fact that a generation of humans and algorithms are learning together to look at images?" (Malevé 2019). His intervention is to ask to what extent machine learning and radical pedagogy might learn from each other.[14] A double movement between teaching and learning can be detected between humans and machines that require a new form of visual learning beyond mere equivalences, and with profound political consequences based on a lack of reciprocity. As previously described, the dataset can be understood to be the algorithm's worldview, which raises the question of what this worldview is, how it was formed, and more importantly how it can be transformed. How might other ways of training elicit other ways of seeing, and in ways that are socially transformative?

By extension, how might machine learning contribute to the transformation not only of learning but also of literacy? How is visual literacy challenged by machine vision and what they learn from each other? If we want to *see* the invisible world of machinic visual culture, we need to unlearn how to see like humans and learn to see more like machines, or rather, see like both in ways that departs from a Western-centred humanist standpoint and thereby embrace intersectional methodologies. The point, drawing from the posthumanities, is that more diversity might elicit a more sympathetic vision of the world – one that is less based on extraction and takes better account of the environment, especially given how resource-heavy machine learning is, as well as other species and how their distinctive ways of seeing expose alternative epistemologies.

What is rendered visible and invisible to perception is clearly crucial to this, especially in a situation where the visual field is increasingly nonhuman and distributed across different entities. Adrian Mackenzie and Anna Munster refer to an "operationalisation" of visuality, in which images operate within a field of "distributed invisuality" in which relations between images count more than their indexicality or iconicity (or aura) of a single image (2019, p. 16). Seeing, or what they call "platform seeing," becomes distributed through data practices and machinic assemblages, that

> emphasize the importance of the formatting of image ensembles as datasets across contemporary data practices; the incorporation of platforms into hardware in devices; forms of parallel computation; and the computational architectures of contemporary artificial intelligence. These assemblages

constitute the (nonhuman) activities of perception as mode of cutting into/ selecting out of the entire flux of image-ensemble world.

Mackenzie and Munster 2019, p. 3

The significance is that a new distributed (or networked) mode of perception is operationalised, what they call "invisual perception" – a new way of (machine) seeing which is an assemblage of its various parts: including imaging devices (such as cameras), the data they produce (which might take the form of an image), and the wider practices and infrastructures through which they are operationalised (in terms of its application) (Mackenzie & Munster 2019, p. 4). So, what kind of literacy is required for such an assemblage that extends literacy beyond representational modes and human sense-making, and is attentive to the relational operations of algorithms, datasets, and infrastructures?

It's clearly not as simple as learning a different vocabulary (although that would help too) but of developing a literacy that is co-constituted, one that is more sensitised to relational operations and that shifts our attention away from the acquisition of technical know-how alone to new possibilities for aesthetic practice – in other words, that helps to expose the politics of invisuality and the envisioning of other potentials. Rather than disregard literacy, or consign it to history, or the false needs of policy-makers, we need it now more than ever to understand how forms of privilege are reproduced and naturalised through new ways of seeing.

Notes

1 *Ways of Seeing* was a documentary made for the BBC in 1972, a four-part television series of 30-min films created by Berger and producer Mike Dibb, and the scripts were adapted into the book of the same name, published by Penguin also in 1972 (see Figure 5.1). Produced at a time when cultural studies was emerging as a field, it was also a repost to Kenneth Clark's BBC television documentary *Civilisation* (1969) which espoused traditional notions of Western culture without political contextualisation. In contrast, *Ways of Seeing* emphasised the politics of representation, and how particular elite forms of knowledge were legitimated to support class, gender, and racial privilege.

2 Transcribed from the last part of the first episode of *Ways of Seeing* (BBC, 1972). Close attention to the means of production, and its alienating effects, further underscores the Marxist approach of Berger, and how exposing what remains hidden allows a fuller understanding of lived reality/conditions. The Brechtian techniques (alienation-effect) used for the television programmes further stress the technical apparatus of the studio, encouraging viewers not to simply watch but also be forced into an analysis of their alienation.

3 This question, how the relation between what we see and what we know is further unsettled by developments in machine vision, has been explored in numerous collaborative events since 2016, organised by the Cambridge Digital Humanities Network, and convened by Anne Alexander, Alan Blackwell, Geoff Cox, and Leo Impett. Also see "Introduction to Ways of Machine Seeing" (Azar, Cox, & Impett 2021).

4 The image was part of Trevor Paglen's exhibition "From 'Apple' to 'Anomaly'" at the Barbican Centre, London, 26 September 2019 to 16 February 2020; also reproduced in Kate Crawford and Trevor Paglen's essay "Excavating AI: The Politics of Images in Machine Learning Training Sets" (2019).

5 By the shorthand "artwork essay," I refer to Walter Benjamin's much cited 1936 essay usually translated as "The Work of Art in the Age of Mechanical Reproduction," that has been a standard reference for analysis of the interrelation of political, technological, and artistic development under capitalism (Benjamin 2002).

6 By "hegemony," reference is made to one of the central concepts of the cultural studies movement in the 1970s, and the work of Antonio Gramsci, to describe how the ruling class manipulates the culture of that society to naturalise their own worldview as the cultural norm.

7 For more on the application of hegemony to computer vision, see Gabriel Pereira's "Towards Refusing as a Critical Technical Practice: Struggling with Hegemonic Computer Vision" (2021).

8 Hoggart's main argument is about the break-up of the old, class culture, lamenting the loss of the close-knit communities and their replacement by the emerging manufactured mass culture at that time (2009).

9 To give some examples: Early chatbots like *Eliza* (created by Joseph Weizenbaum at MIT between 1964 and 1966) follow similar principles. It is named after George Bernard Shaw 1913 play *Pygmalion* (loosely based on the Greek myth) describes a bet by a professor of phonetics that he can teach a working class girl, Eliza Doolittle, upward mobility through the British class system, through the acquisition of "proper speech" (as opposed to Cockney dialect). A further example is Sarah Ciston's chatbot *ladymouth* (2015), a project that tries to explain feminism to misogynists on Reddit. A further explanation can be found at https://gitlab.com/sarahciston/book/-/tree/main/source/8.5-TalkingB ack#fn3-4211.

10 https://esolangs.org/wiki/Brainfuck.

11 The wider discussion of decolonial computing resonates with this, such as the work of Syed Mustafa Ali (2016) who argues that computing is founded upon aspects of colonialism, and that contemporary developments point to an intensification of the colonial impulse.

12 Benjamin's notion of "optical unconscious" and what others have called the "machinic unconscious" are invoked by this.

13 These phenomenological and semiotic references find synthesis in a post-phenomenological approach of Bernard Stiegler, whose work recognises the originarity of technology, explored in more detail in "Introduction to Ways of Machine Seeing" (Azar, Cox, & Impett 2021).

14 This question, and much of the preceding description, are based on a current research project that builds on both Nicolas Malevé's research and the ongoing work with Cambridge Digital Humanities from 2016 (see endnote 3).

Bibliography

Ali, Syed Mustafa. 2016. "A Brief Introduction to Decolonial Computing." *XRDS: Crossroads, The ACM Magazine for Students* 22 (4): 16–21.

Azar, Mitra, Geoff Cox, and Leo Impett. 2021. "Introduction to Ways of Machine Seeing." *AI & Society* (special issue). https://doi.org/10.1007/s00146-020-01124-6.

Benjamin, Walter. 2002 [1936]. "The Work of Art in the Age of Its Technological Reproducibility: Second Version." In *Selected Writings, Volume 3, 1935–1938*, edited by Howard Eiland and Michael W. Jennings. Cambridge, MA: Belknap Press of Harvard University Press, pp 19–55.

Berger, John. 1972. *Ways of Seeing*. London: Penguin.

Berger, John and Mike Dibb. 1972. "Camera and Painting." *Ways of Seeing*. Episode 1. BBC Two.

Celiński, Piotr. 2019. "Literacies." *Media Linguistics* 6, (4): 467.

Ciston, Sarah. 2019. "Ladymouth: Anti-Social-Media Art as Research." *Ada: A Journal of Gender, New Media, and Technology*, (15). https://doi.org/10.5399/uo/ada.2019.15.5.

Crawford, Kate and Trevor Paglen. 2019. "Excavating AI: The Politics of Images in Machine Learning Training Sets." https://excavating.ai.

Eglash, Ron. 2007. "Broken Metaphor: The Master-Slave Analogy in Technical Literature." *Technology and Culture* 48 (2) (April): 360–369.

Farocki, Harun. 2004. "Phantom Images." *Public: New Localities*, (29). https://public.journals. yorku.ca/index.php/public/article/view/30354/27882.

Hall, Stuart. 1980 [1973]. "Encoding, Decoding." In *Culture, Media, Language. Working Papers in Cultural Studies, 1972-1979*, edited by Centre for Contemporary Cultural Studies, London: Routledge, pp 128–138.

Harney, Stefano and Fred Moten. 2013. *The Undercommons*. Wivenhoe/New York/Port Watson, NY: Minor Compositions/Autonomedia.

Harwood. 2008. "Class Library." In *Software Studies: A Lexicon*, edited by Matthew Fuller. Cambridge, MA: MIT Press, pp 37–39.

Hoggart, Richard. 2009 [1957]. *The Uses of Literacy: Aspects of Working Class Life*. London: Penguin.

Knuth, Donald E. 1984. "Literate Programming." *The Computer Journal. British Computer Society* 27 (2): 97–111. https://doi.org/10.1093/comjnl/27.2.97.

Mackenzie, Adrian and Anna Munster. 2019. "Platform Seeing: Image Ensembles and Their Invisualities." *Theory, Culture & Society* 26: 3–22. https://doi.org/10.1177/0263276419847508

Magritte, René. 1927. *The Interpretation of Dreams (La Clef des songes)*. Painting. Art Institute of Chicago.

Magritte, René. 1929. *The Treachery of Images (Ceci n'est pas une pipe)*. Painting. Los Angeles County Museum of Art.

Malevé, Nicolas. 2019. "An Introduction to Image Datasets." *Unthinking Photography: The Photographers' Gallery*. Available at https://unthinking.photography/articles/an-introduct ion-to-image-datasets.

Merleau-Ponty, Maurice. 1968. *The Visible and the Invisible*. North Western University Press. Available at https://monoskop.org/images/8/80/Merleau_Ponty_Maurice_The_Visible _and_the_Invisible_1968.pdf.

Mirzoeff, Nicolas. 2011. "The Right to Look." *Critical Inquiry* 37 (3) (Spring): 473–496.

Paglen, Trevor. 2019. *The Treachery of Object Recognition*. Print. "From 'Apple' to 'Anomaly'," exhibition, Barbican Centre, London, 26 September 2019 to 16 February 2020.

Pereira, Gabriel. 2021. "Towards Refusing as a Critical Technical Practice: Struggling with Hegemonic Computer Vision." *APRJA* 10 (1). https://doi.org/10.7146/aprja. v10i1.128185.

Vee, Annette. 2017. *Coding Literacy: How Computer Programming Is Changing Writing*. Cambridge, MA: MIT Press.

Virilio, Paul. 1994. *The Vision Machine*. Bloomington, IN: Indiana University Press.

Wing, Jeannette. 2006. "Computational Thinking." *Communications of the ACM* 49 (3): 33–35.

6

SOFT SUBJECTS

Hybrid Labour in Media Software

Alan Warburton

> The good practitioner needs to be able to reflect upon their own context. To
> see work and employment as having a context is to get nearer to being able
> to scrutinise and reflect upon the conditions of production.
>
> Dewdney and Ride 2014, p. 150

As befitting a good practitioner contributing to a book about networked images,
the subject of my interest is the experience of *people who make images using soft-*
ware, at first glance an impossibly broad category that covers most forms of contemporary visual culture: artists, filmmakers, architects, designers, animators and
technicians working across art, film, advertising, product design, fashion, virtual
reality and countless other industries and niches. But there's good reason to pull
this disparate group out of the multitude and address them together: they are
already joined by (and joined *with*) their tools. While each industry might have
their own specialist software packages – fashion designers use Clo3D, architects
use Autodesk Revit, visual effects technicians use SideFX Houdini – they also collectively circulate and converge around common hardware, software ecosystems
and applications, common workflows, file formats and technical rationales. As
such, their lives and their labour are individuated through software; their brains
moulded to common software logics, their hands accustomed to common key-
board shortcuts and their livelihoods dependant on and synchronised to the same
cloud subscriptions and updates. Based on this commonality of workers and
machines, this chapter argues for a distinct status and recognition to be afforded
to this kind of media worker.

The challenge, however, is that these individuals remain largely divided from
each other due to the persistent legacies of analogue specialism, academic or
industrial. While some progress is made within their respective industries (in trade
press, professional networks, product reviews, industry forums or events) towards

DOI: 10.4324/9781003095019-9

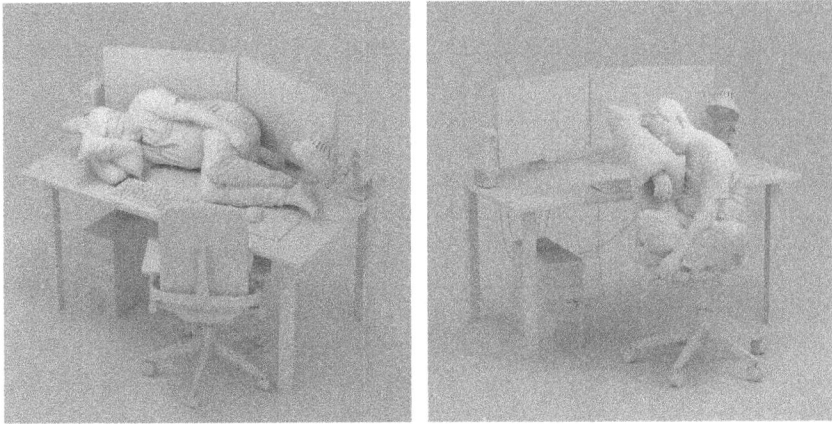

FIGURE 6.1 Alan Warburton, *Sprites I-IV* (2016)

acknowledging computational capitalism's mode of production, critical discourse does not often broach the walls of the silo. Common to most digital industries, however, is a habitual focus on innovation and creativity in lieu of attention on political, social or ethical questions. This is despite the fact that the animation, film, architecture, design, visual effects and games industries are at the frontline of many of the most controversial technical advances in image production, which usually involve machine learning, image recognition and image synthesis. They are also longstanding hotbeds of toxic work practices (Cote and Harris 2021, p. 3) and are increasingly being restructured by automation. Even as schedules tighten, wages drop and 'innovation' renders their specialist labour obsolete, precarious and 'prosocial' media producers are often precluded from unionising or even speaking out (Hesmondhalgh and Baker 2008, p. 104), unlike the defecting tech CEOs who publicly profit from their disillusionment with Silicon Valley. This double bind of complicity is compounded by the fact that when your livelihood depends on being 'inside' software and 'making cool stuff,' the politics of media work often become necessarily subordinate to the challenge of marshalling constantly changing technical systems to produce spectacular or novel creative artefacts.

McKenzie Wark's concept of the 'hacker' (for her, this means anybody who produces new value through the manipulation of data) describes the pressures of this kind of work:

> [t]he workplace nightmare of the worker is having to make the same thing, over and over, against the pressure of the clock; the workplace nightmare of the hacker is to produce different things, over and over, against the pressure of the clock.

2019, p. 29

Here we begin to understand that developing mastery over cutting-edge software tools and participating in a broader culture of creative innovation propels many practitioners forward, yet in the end, it can drain their creative problem-solving energy. This is an effect described well by Software Studies scholar Matthew Fuller as the exploitation of capacity of the 'general intellect' to 'take bits from here and from there, to recompose multiply encoded and gated, broken, esoteric and public materials and information and make something of them' (2006, p. 9).

In a volume on software's role in contemporary architectural practice, architect Galo Canizares illustrates a similar point, arguing that one of the principal tasks of the digital architect is now to 'translate across a vast, ever-updating landscape of standardized file-types and graphical user interfaces, simultaneously making decisions about users, permissions, and longevity' (Canizares 2019, p. 523). As a practitioner with 15 years of professional experience in various industries, this is one of very few published accounts of media labour that begins to approach the realities of practice. Most of my time is spent troubleshooting renders, versioning and resolving technical incompatibilities; sourcing, repairing, formatting and customising images, 3D models, templates, databases, systems and archives; installing and applying software tools, drivers, plugins, hotfixes, patches and beta updates; researching error messages, posting on forums, reading product documentation and searching for tutorials; downloading ancient freeware, cracks and torrents using VPNs, searching terabytes of hard drive data, pulling favours from friends, bartering with render farms or (as a last resort) outsourcing to gig-work platforms. All of this activity is shaped by high-pressure schedules that crystallise my labour into ever more dense and economical forms, forcing me to find efficient technical workarounds that allow me to create something bespoke (but rarely profitable) from systems only partially designed for my purpose. I am one of those people – like Canizares' architect – that spends as much time in the dysfunctional in-between spaces of software than I do in the walled gardens of Adobe and Autodesk's flagship applications.

For busy practitioners like us, the machinations of technocapital are primarily felt through the various frictions in our tools and workflows, often spurred on by the incremental improvement in graphics processing units (GPUs). A practitioner might well ask themselves (while waiting for a GPU driver update to install or a render to finish, perhaps) where this endless parade of improvement and innovation is heading – what's it all for? Orienting practice in a complex and ever-changing landscape of technical evolution is challenging, and for the practitioner it can feel like trying to find a path home in the morphing fractal cities of *Inception* or *Doctor Strange*. This is the kind of work, the kind of software, and the kind of collective subject that I'm interested in. *Who are we? Where are we? What are we doing here?*

Answering these questions should be as useful for the practitioner as the academician, and this chapter attempts to bridge the interests of both. The goal in the first instance is not to rush into a revision of existing critical software 'lexicons,' 'tactical media' strategies or models of political resistance. Instead, I intend to attend closely to the inescapable practices and politics of work within today's sophisticated media ecologies, to make more legible the phenomenological experience of professional media producers whose practices are tightly defined by the affordances of mainstream software applications and ecosystems (whether commercial or open-source). I intend to build a picture of media practice useful for scholars of media, but also for practitioners themselves, who I hope are able to recognise themselves in this picture and use it as a guide to orient themselves within the unpredictable dynamics of media software evolution.

My hypothesis is that the practitioner I'm interested in is a distinct sociotechnical entity whose labour and agency is defined by common apparatus, affordances, working practices and relationships to media technology. To describe this entity simply, I propose the placeholder term 'soft subject.' This term introduces a few covalent concepts: *soft* of course implies that these individuals from different industries and disciplines are joined by software tools and practices, but also that their inherent interdisciplinarity renders them fuzzy with overlaps. Softness also conveys pliancy, which refers to the fact that digital media, techniques, materials, labour markets and workers are constantly mutating (because new media, of course, continues to be new).

I'll attempt to bring more definition to this hypothetical soft subject by drawing on existing critical concepts from a variety of disciplines. At the broadest level, the soft subject reflects general trends within contemporary post-Fordist labour discussed at length in media studies, sociology, cultural studies and political economy. Distinctions between immaterial labour, affective labour, cognitive labour and creative labour help define soft subjectivity, but for further resolution, fields like labour process studies and production studies attend to the technical and organisational qualities of media software practice in more detail. While all these fields provide useful routes into understanding the specificity of the soft subject's labour, if that labour is at least in part structured by *software ecosystems*, then deeper studies of software politics and software affordance – the ways software might 'request, demand, allow, encourage, discourage, and refuse' a subject's freedoms (Davis and Chouinard 2016, p. 242) – are of equal value. I'll draw on media studies, new media theory and media archaeology here, but particularly useful is the interdisciplinary field of software studies, which emerged from the early internet cultures of the 1990s and often formulated critical strategies to deal with hegemonic software 'sensoriums' (Fuller, 2006, p. 15). This aspect of software studies remains instructive for critical coders, but adjustments are needed for a media practitioner who makes their living from the use of powerful commercial software applications: even if software grinds them down, 'hacking' a 'speculative' alternative is beyond their means, as it would be even for most software developers. This necessarily antagonistic complicity with the increasingly powerful affordances of commercial software is the starting point for many other software studies scholars, however. Gehl and Bell's investigation into the industrial cultures of software design explores how technological, discursive and human elements are 'associated' by the 'disciplined' software engineer into cohesive artefacts that, in turn, are 'disassociated' in their deployment by 'unruly' users (Gehl and Bell 2012, p. 15). This describes well the way that software users both rely on suites of automated functions but also seek to push beyond prescriptive outputs in the pursuit of unique results. Adrian Mackenzie digs further into this mode of labour when he suggests that the work of engaging with machine learning software is 'not so much implementation of particular techniques [...] but rather navigating the maze of methods and variations that might be relevant to a particular situation' (Mackenzie 2017, pp. 75–76). Both of these perspectives chime with the description of software labour I've already provided, suggesting that the sociotechnical work of the soft subject might be thought of as a blend of creative, strategic and technical work that is enabled by automation but also at odds with it.

For a detailed assessment of media software that blends a lucid analysis of culture, labour *and* software, the work of new media theorist Lev Manovich is particularly useful. It is for this reason that in this chapter I'll test ideas from Manovich's influential *Software Takes Command* (2013) against other scholarly concepts and current practice to develop a more precise understanding of my speculative 'soft subject.' My argument will proceed in two main sections that answer two overarching questions. The first question, 'where does the soft subject work?' will seek to locate and describe the sites of practice in which media workers of multiple disciplines converge. I will focus on the ways Manovich uses his ontology of software media to offer a formal account of divisions and classifications within software ecosystems and consider whether these theoretical classifications help to locate the practices of the soft subject. The second question 'how does the soft subject work?' will build on the conclusions of the first, using Manovich as a starting point to understand the consistent ways that software automation structures labour. Synthesising responses to both questions, my conclusion will reflect on whether a stable concept of 'soft subjectivity' can be articulated, and if so, what the wider implications of this might be.

In *Software Takes Command*, Manovich provides a legible guide to software and its expression as computational ontology, an aesthetic medium and a condition of cultural production. Manovich's main idea comes from a revision of Alan Kay and Adele Goldberg's concept of metamedia, where he suggests that computers don't simply *aggregate* media simulations, they *hybridise* them. This is a result of the 'many to one mapping from physical materials to data structures' (2013, p. 204) that leads to extensive interoperability, whereby 'different media are actively trying to reach towards each other, exchanging properties and letting each other borrow their unique features' (p. 65). Software users are characterised as being able to 'combine elements created in different software applications, or move the whole project from one application to the next to take advantage of their unique possibilities' (p. 247).

Manovich's analysis of media hybridity attends to practice in ways that other theories of digital remediation do not. Yet for critics of Manovich, he doesn't go far enough in this direction. Media theorist Alexander Galloway suggests (2012,

p. 24) that Manovich's focus on an essential software ontology falsely begins with 'objects and operations' over the more apposite 'practices and effects.' This might also be thought of as a framing problem: Galloway thinks that Manovich finds the conditions of metamedia within the computer, rather than in an 'ethic' of interaction or practice. Galloway is looking for a more political, sociotechnical account of how computers define 'a practice or a set of executions or actions in relation to a world' (p. 19). In Manovich's defence, many of the concepts that arise from his investigation of hybridity are fundamentally processual: 'soft evolution,' 'deep remixability' and 'permanent extendibility' all describe conditions of action and change, and are useful as starting points for soft subjects whose precarious labour is shaped by a recursive process of disruptive innovation. Galloway's point stands, however: while Manovich does an excellent job of triangulating between ontology, culture and creative practice, it is the third area – the sociotechnical 'ethic' of metamedia – that remains latent in its development. A decade (or two) on from his key texts, there may therefore be room for a 'remix' of hybridity that uses knowledge of contemporary software practice to locate an *identity* and an *ethic* for the natively new media hybrid practitioner that I term the soft subject. So, how does Manovich's manifesto manifest? *Where* and *how* does the soft subject work?

Where Does the Soft Subject Work?

Locating the site of hybrid media practice requires an understanding of how individuals from diverse industries and media backgrounds converge in the use of certain software applications. This 'many to one mapping' (2013, p. 204) of old media to new computational categories is central to Manovich's new media ontology, but does he offer any kind of topological mechanism to identify where the soft subject might be located? The mapping tools he provides us with accord to a variety of nested ontological parameters: software provides a variety of *algorithmic operations* (old media vs. new media, media independent vs. media specific, and so on) and *data types* (text, vector graphics, raster graphics, 3D models, video). These are incisive, essential concepts, and while useful for identifying media DNA, Manovich himself acknowledges the limitations of his categorisation system when he says that his distinctions work better in theory than in practice (2013, p. 112). The reason for this is that according to the principle of hybridity, if any program can theoretically accept any data type and adopt almost any algorithmic operation and combine these in any permutation, then we're as likely to see applications that are *old media* specific (photography), *new media* specific (based around data types like raster images, or file formats like .jpg or .png), *task/algorithm specific* (photographic grading) or *workflow* specific (processing, grading and publishing of photographs). Indeed, all these options (and infinite variations therein) are present in today's media landscapes, but that doesn't account for the fact that, in practice, most media production revolves around a relatively narrow range of popular software applications with specific affordances. How can this aspect of media-in-practice be accounted for? Manovich

provides an explanation that concisely demonstrates where the bounds of his ontological theory of media lie:

> Various social and economic factors—such as the dominance of the media
> software market by a handful of companies or the wide adoption of par-
> ticular file formats — also constrain possible directions of software evolution.
> Put differently, today software development is an industry and as such it is
> constantly balancing between stability and innovation, standardization and
> exploration of new possibilities.
>
> *Manovich 2013, p. 93*

This brief outline of the industrial forces that shape the continual expansion of
software understates an important complexity that many would consider essential
to any media mapping exercise. Manovich's preference for the ontological over the
social, political or industrial precludes him from effectively providing this. Even
when offering a concrete conceptual framework to describe the 'soft evolution'
of media ecosystems, he cleaves to a notion of infinite speciation that disavows
the possibility of any convergent software 'monomedia' as being contrary to the
principles of endless media hybridity (2013, p. 171). This might be ontologically
sound, but it's not a useful way of identifying the patterns of software use and
development seen 'in the wild.' Rectifying this at the most obvious level, I would
suggest that the majority of software users – including the soft subject – circulate
around commercial software ecosystems, primarily those developed by Adobe and
Autodesk.

Elsewhere, Manovich inadvertently provides another semi-stable classificatory
distinction, but doesn't position it explicitly as such: in his discussion of motion
graphics, which he goes so far as to suggest is 'as important historically as the inven-
tion of printing, photography, or the Internet' (2013, p. 250), he attends to the ways
that some new media applications are more 'revolutionary' in their hybridity than
others. Programs like After Effects allow for exceptionally high degrees of interoperability through the combination of multiple algorithmic tools, media paradigms,
data types and cultural outputs. This is an 'avant-garde' kind of 'deep remixability'
(2013, p. 248), where new media objects are *composited*, or 'put together from elements which come from different sources' and those elements are 'coordinated and
adjusted to fit together' (Manovich 2002, p. 139). His discussion of After Effects
tends carefully towards the aesthetic language of hybridity and less to practice, but he
does acknowledge that motion graphics software is where the 'professional boundaries between different design fields' (2013, p. 247) are dissolved and individuals
and media objects from multiple industries, media traditions and labour specialisms
overlap. Motion graphics might therefore be thought of as an acute example of the
interdisciplinarity that characterises media hybridity, and as such prompts a further
distinction: 'hybrid' soft subjects converge in the use of *generalist* and highly *interoperable* software used by multiple industries and workflows, as opposed to within the
specialist software used within specific industries and workflows.

What's interesting about these generalist applications is that most are 3D animation, compositing, visual effects and games design programs that are notable for accepting a wide range of file formats and data types. This makes sense intuitively: an application that operates according to multiple 'visual, spatial, and temporal' logics (2013, p. 249) is like a universal adapter that accepts multiple inputs, whereas applications designed to manipulate images only through 2D interface logics (layer stacks, for example) are more 'regional.' The centrality of this acutely 'new' media paradigm and its continued evolution is reflected in cultures of professional practice: whereas ten years ago students and professionals alike migrated en masse to accessible 3D modelling and compositing software like Cinema 4D or After Effects, focus has since shifted to games engines like Unity and Unreal. Unity, particularly, is a site of significant convergence, used by the games and film industries, but also by BMW, Chinese social media giant Tencent and construction firm Skanska (Peckham 2019). Large global hardware and software companies are increasingly invested in and reliant on these interactive 3D applications: in 2021, Nvidia rolled out the ambitious *Omniverse*, a web-based production engine which offers interoperable design, simulation and rendering tools; Facebook pivoted their entire business towards the 'metaverse' of virtual reality (VR) and augmented reality (AR) communication; the development fund of open-source 3D application Blender attracted the investment of Apple on top of the likes of Intel, AMD and Google, and Unity's stock market IPO saw the company valued at over $1.3 billion.

What can account for the increasing importance of these generalist 'worldbuilding' applications? In the essay *Image Future* (Manovich 2006), a version of which was cut from *Software Takes Command*, Manovich suggests something important: natively new media software applications are particularly notable for the fact that they introduce standards of *editability* and *configurability* that increasingly condition all forms of production. Building on ideas from *The Language of New Media* regarding 'modularity' (Manovich 2002, p. 137), Manovich helps us see what the invasive hybrid force of new media is really characterised by: the modular logic of flexible new media 'assets' (3D objects made of vertices and faces, images made of pixels, pixels made of code) has set a new baseline for all media production processes. In *Image Future*, Manovich discusses how digital apparatus allows the fusion of analogue and digital through 'universal capture' methods (2006, p. 3) that sample the real world in discrete ways that befit digital production methods. This forced conversion of relatively inflexible analogue wholes to *flexible, editable* and *configurable* digital parts seems to be the evolutionary survival principle of metamedia, the superpower that determines why new media 'computational' ontologies overtake old media 'cultural' categories, why they continually produce change within software ecosystems and supply a sequence of new standardised file formats, applications, plugins, devices and workflows. Continual change in software practice occurs (in part) through the introduction of ever more modular, precise and flexible methods of modelling images. This is perhaps the underlying driver of the 'innovation, standardization and exploration of new possibilities' (2013, p. 93) that Manovich fails to really elucidate in his description of the 'social and economic' factors of software evolution.

Other scholars help clarify how this tendency is actually an expression of a deeper epistemological move towards increased *control* (via capture, modelling and simulation). Alexander Galloway (2012, p. 19) says that 'computers themselves are those special machines that nominalize the world, that define and model its behavior using variables and functions' and building on this, media archaeologist Jacob Gaboury (2021, p. 185) suggests that computation is a 'recursive process oriented toward the simulation of the world as discrete computational objects'. Media theorist Bernard Dionysius Geoghegan (2019, p. 87) looks at how control is embodied by interactive screen interfaces that correspond to a military-industrial logic of 'regulating, enlisting, breaking-down of space, mapping', and image theorists Ingrid Hoelzl and Rémi Marie apply a similar philosophy of mapping and control to the digital image specifically, introducing the concept of the operational 'softimage' which they characterise as symptomatic of an epistemological shift to a 'flexible image of a flexible world' (2018, p. 134).

In these critical perspectives, we can see that manifestations of 'innovation' in imaging technologies ('improved' algorithms, devices, software, interfaces and images) actually express an epistemological drive towards increased mastery, flexibility and modularity through media.[1] The evolving 'worldbuilding' software package exemplifies this recursive process, and the hybrids that operate such software are the vanguard of *control*. For better or worse, this is where the soft subject works: at the portable, immersive, real-time frontline of software evolution, in the generalist applications of industrial software ecosystems, as a media 'avant-garde' battalion marshalling *control* over simulation technologies.

How Does the Soft Subject Work?

If the marshalling of control through media software is the broadest way of characterising the work of the soft subject, how is this achieved in practice? Manovich helps open up ways of thinking here by suggesting a fundamental distinction between two kinds of tool the software user interacts with: 'low-level' and 'high-level' (2013, pp. 128–129) algorithmic automation. 'Low-level' automation manifests in software tools that require a lot of manual control: a digital paintbrush tool that allows for detailed user input, for example. 'High-level' automation, on the other hand, involves algorithmic tools that condense multiple complex functions together, often using artificial intelligence: an automatic skin smoothing filter, for example. Manovich pegs low-level and high-level automation to industrial and consumer applications respectively, suggesting that 'if we equate the use of computers with automation, paradoxically it is the consumers who fully enjoy its benefits—in contrast to professionals who have to labor over all these manual settings of all these controls' (2013, p. 224). Here he's invoking the way consumer smartphone apps rely heavily on 'high-level' filters whereas professional, industrial applications have granular 'low-level' controls.

Many media practitioners would quickly identify a problem here. In his discussion of 'template culture' within his industry, graphic designer and media scholar Silvio Lorusso does just this, noting how automated graphic design websites that 'make use of neural networks and therefore have the ability to improve their designs by "learning" from their mistakes' (2017, p. 1), have led to the automation of much graphic design work and fundamentally altered the labour of the graphic designer, who must now pivot to intellectual activities like writing or research. This automation of professional creative labour comes through what Manovich would consider to be the high-level automation found in consumer media applications, yet clearly for Lorusso and many others, it has significant effect on professional practice. It's evident, therefore, that high-level automation regularly occurs throughout professional media ecosystems, often rendering specialist labour obsolete through advances in hardware power and artificial intelligence.[2]

Manovich himself doesn't directly address this, but classical Marxist thought suggests that in the inevitable historical progression from craft production to small-scale manufacturing and then to industrial production, the skill of the worker is progressively automated and subdivided, to the benefit of the owners of factories, machines (and software). Scholars in the field of labour process studies, however, have contested this deskilling narrative. Paul Adler suggests instead that instances of deskilling are 'eddies in the broader current of a long-term skill-upgrading trend' (1990, p. 783), and in a study on the 'micropolitics' of digital skills, geographers Richardson and Bissell clarify Adler's view, suggesting that 'the intricacies of technology design and operation require development of new skills; and it is these new skills that can be overlooked in arguments that point to deskilling tendencies' (2019, p. 282).

So, does software instigate a progressive deskilling of human labour through the introduction of high-level automation, or does it continually require an equal degree of upskilling and control over new forms of low-level automation? I would suggest – further to my conclusion about the expanding frontier of control in media software – that in distinction to Lorusso's specialist, the mobile soft subject is characterised by the second option. They progressively migrate towards a frontline of new media software that is not only *epistemologically* and *ontologically* more totalising in its control, but that *practically* involves more manual operation. A good example of this is in the accelerated development of real-time graphics in games, AR, and VR. VR, despite numerous false starts over the past decades, is currently at the complex frontline of technical evolution in image-making practice, and it's for this reason that it will provide a key final analysis of the soft subject.

VR experiences are produced entirely in 3D, must run faster and smoother and at exponentially higher resolutions than any other visual media format, all while conforming to increasingly mobile form factors. This is an exciting, cutting-edge production environment where the buzz of innovation is constantly felt through collectively shared 'wow moments,' registered when artists increase believability, immersion or photorealism despite the technical obstacles to doing so. However, the cognitive, affective and technical overheads borne by the artist in pursuit of this goal should not be underestimated.

VR artists are soft subjects par excellence, primarily because they are not working within stabilised, mature ecosystems with a lot of standardised high-level automated controls, but with emerging ecosystems where there is a lot of contingency, improvisation and low-level control. The tools used in production are classically *generalist* worldbuilding applications. Modelling and mesh optimisation often happens in programs like Maya, Photoshop, Cinema 4D, Blender or 3D Studio Max, but animation, interaction and lighting usually come together in game engines like Unity and Epic's Unreal, which are sites of rapid technical development.

One of the most significant and laborious 'low-level' processes involved in VR production is 'optimisation.' VR artists optimise by constantly trying to manage the size of texture files and geometry, the complexity of interaction or the details in lighting calculations. This assertion of control aims for multiple moving targets: competing consoles and headsets that rapidly iterate various formal innovations in design, processing power, fidelity and immersion. Optimisation for multiple platforms is a necessity for most VR studios because releasing a product on only one wouldn't be enough to recoup production costs. Artists must therefore bridge the disparate technical capabilities of powerful high-end consoles like PlayStation and economical mobile VR platforms developed by Samsung or Google, finding the computational 'sweet spot' for each platform. This requires a detailed understanding of multiple rapidly evolving API specifications and the way each allocates memory ('draw calls') and processing power. The VR artist is therefore constantly switching between desktops and headsets and phones to create, compile, test and export content for different platforms and devices.

What this means in practice is that the sociotechnical skills of the VR artist consist of tracking and evaluating the relative technical affordances of many devices, algorithms, file formats, workflows, APIs, renderers, plugins, applications and platforms. Do you begin a job with low-level controls that maintain flexibility and modularity but require slow manual operation? Or rely on the high-level presets that don't quite work but get the job done quicker? Which workflow is more stable for which platform? Do you optimise for current builds or wait for the next hardware release or software update? Do you hire a coder to customise your tools, post on a forum, request a bugfix from developers or learn to code yourself? Do you have time to assess these options? Unfortunately, in media cultures that equate work with passion and naturalise overwork, these nuanced sociotechnical decision-making processes are simply considered part of 'getting the job done.' What should be clear, however, is that as soft subjects par excellence, VR artists don't fit easily into classical models of industrialised labour or narratives of progressive deskilling. They operate in a complex sociotechnical milieu that incorporates aspects of bespoke craft production, small-scale manufacturing and full-scale industry. It may be helpful, therefore, to apply critical concepts provided by other scholars to more precisely describe this kind of labour and propose an answer to the question 'how do soft subjects work?'

Production studies scholars Mark Deuze and Mirjam Prenger helpfully characterise the effects of media hybridity on makers of media, who act as a buffer between media organisations and unstable media ecosystems, becoming 'simultaneously generalists and specialists, combining the command of one profession with knowledge of a host of others' (2019, p. 20). This concept is particularly useful because it helps clarify how Manovich's concept of media hybridity might be transposed from a condition of computational media to a condition of practice, or what Alexander Galloway called for as the missing 'ethic' of media hybridity. I would suggest that this 'hybrid media ethic' could be clarified further: if interpreted according to classical Marxist ideas of industrial production, hybrid soft subjects are neither blue-collar stationed specialists nor white-collar generalist process managers, but *multispecialists*. As the VR artist demonstrates, their work involves operating across multiple roles, with both high-level and low-level automation, with multiple data types and algorithmic techniques, facilitating the translation of data between many competing applications and platforms.

Elaborating on this ethic, we can see that what compounds the challenges of the soft subject's *multi-specialism* (and renders it anathema to deskilling narratives) is that it occurs within emergent and competing hardware ecosystems that outpace the ability of software developers to design efficient tools and workflows. This means artists, studios and creative communities are not simply upskilling by learning new low-level tools, but are themselves creating those tools, assembling complex constellations of hacked-together scripts and plugins that bypass limitations, increase control or shorten repetitive or unstable workflows. This contradicts the assumptions that many might have that working at the frontline of imaging technology simply involves 'porting' previous innovations to newer, smaller, faster hardware platforms. In fact, as we see in the importance of VR optimisation, working with emergent platforms involves the creation of entirely new workflows, creative strategies and algorithmic techniques that attempt to match or exceed the standards of fidelity, immersion and photorealism that prior platforms provided.

The recursive upskilling inherent in the soft subject's 'ethic' can be thought of as a constant attempt to 'habituate,' a concept explored by media theorist Wendy Chun who suggests that the constant processes of innovation and obsolescence in commercial hardware and software systems 'deprives habit of its ability to habituate' (Chun 2016, p. 97). With software seemingly always in beta and hardware quickly obsolete, users must accept as natural the constant economic and cognitive overheads of what Chun would term 'updating to remain the same.' But when this recursive change occurs across *multiple* applications, platforms and ecosystems, attempts to habituate become exponentially more challenging. Soft subjects must constantly overcome the 'path dependency' of technical systems, performing the sociotechnical version of linguistic code-switching. While it's well understood that many who transitioned from old media to new are part of generation that 'acquired new skills and changed their practices "on the job"' (Dewdney and Ride 2014, p. 207), the effects of impeded habituation are multiplied – and endemic – for the soft subject working with new, non-standard, unstable technical systems. It can take weeks or months before a level of fluency is achieved in just *one* new system, and much of that fluency comes from finding stable workflows that avoid triggering memory errors, glitches and crashes. Here we see that the update is far from a linear path to an easier life: within rapidly cycling systems of technical innovation, the challenge is not simply in upskilling and habituating to new control systems, but mitigating the cumulative effects of *technical debt* (Tom et al., 2013). As software studies scholar and coder Adrian Mackenzie observes, '[c]ode is being constantly superseded. The constant arrival of new versions, updates and patches both conceals and highlights the brittleness of software' (2006, p. 12). Rapidly obsolescing technologies pile new functions on top of old, and rarely are systems rewritten.[3] The accumulation of 'bitrot' (Mackenzie 2006 p. 12), 'bloat' or 'feature mountains' (Fuller 2003, p. 140) is not just a reflection of the accumulation of functionality, but happens as a matter of course in software development, driven as it is by the philosophy of 'launch now, fix later.'

So, how does the soft subject work? Using the example of complex VR production, I've proposed that the soft subject could be described as a multispecialist, whose command of low-level automation is a constantly impeded attempt to habituate to emergent software and hardware ecosystems. Building on the conclusions of the previous section, those ecosystems are defined as consisting of generalist worldbuilding applications that express an epistemological move towards a paradigm of flexible, modular media simulation. The labour of soft subjects, therefore, provides the connective tissue in the evolving interstitial spaces of disparate and unstable worldbuilding systems, forging future practices and technical systems that optimise the creation of real time, immersive and photorealistic content for ever-smaller, more portable devices. The labour of the soft subject is creative, certainly, but their real specialism is applying this creativity in the reconciliation of disparity within novel technical assemblages, and to do this they must recursively metabolise obsolescence by pioneering new forms of computational control, overcoming path dependence and habituating to high levels of technical debt.

Conclusion

This chapter has sought to establish, primarily through critical analysis of the work of Lev Manovich, an ethic of new media practice native to *those who make images using software*. While many media professionals would recognise their labour in the description of soft subjectivity I've provided, many might dismiss its problematisation. For some, the complex and demanding nature of this frontline media work is the key attraction of new media creativity. Yet the prevalence of this kind of sociotechnical labour clearly has systemic impacts. In a study of "crunch" in the games industry, Amanda Cote and Brandon Harris found that one of the main factors that contributed to the overwork, fatigue and mental health issues of game industry workers was a constant revision of external technical standards, APIs, devices, platforms and operating systems (2021, p. 2). One significant new platform announcement from Sony or Nintendo could torpedo a development schedule, and a succession of smaller changes across multiple evolving software ecosystems can unpredictable knock-on effects for thousands of workers (something only magnified in the turbulent ecosystems of VR production).[4]

What should be clear from my argument as well as those of the scholars I've cited is that the rapid iteration of technology is a feature of media technology, not a bug. Likewise, 'crunch' is a feature of media work, rather than an avoidable by-product of poorly managed organisations. This is perhaps the reason why job ads in media call for 'VFX wizards,' 'code ninjas' or 'digital gurus': media production depends on multispecialist creative workers whose real value is in their ability to metabolise 'crunch' by mastering and integrating multiple changing software packages, processes and platforms.

This gets at the political heart of the soft subject and helps us focus on the missing 'ethic' of new media Alexander Galloway called for in response to Manovich's ontology. The real importance of the soft subject lies in their relationship to automation and obsolescence: in contrast to the creative media specialist who is rendered progressively obsolete by 'high-level' artificial intelligence, the soft subject is a type of creative worker whose complex multi-specialism (for now) evades this kind of automation. They are the native hybrid artists of new media whose pliable labour prevails past the point of others.

Considered this way, the soft subject is a specific and highly valuable labour resource and a key component of the creative industries within contemporary networked capitalism. The innovative, dedicated artists, technicians, designers and indie studios 'risking it all' to develop content for new ecosystems might even be thought of as the unpaid R&D or marketing divisions of Nvidia, Facebook, Adobe or Autodesk. This consideration of the complex nature of media work has been addressed in various ways: Deuze and Prenger's 'hybrid' is a precariously employed media manager to whom a company outsources risk (2019, p. 20); Matthew Fuller's 'artist' is 'a cheap means to find new uses for new tech' (2006, p. 7) and McKenzie Wark's (2019) 'hacker' is the creative force of digital capitalism whose intellectual property is privatised, packaged, automated and rented back to workers as a subscription service. The soft subject explored in this chapter has something in common with all of these models of creative labour, but has a specific relationship to the operation of contemporary software ecosystems: in their connection to rapidly changing worldbuilding apparatus, their sociotechnical work habituating at the avant-garde frontline of media evolution might be considered, politically, as the gentrification of virtual worlds. Their 'hack' is a form of sociotechnical standardisation 'that gives to production its formal, social, repeatable and reproducible form' (Wark 2004, p. 82), but what is produced by increment is not simply better software, smoother interaction or more photorealistic graphics, but the foundations of a totalising commercial 'metaverse' that will likely seek to capitalise on all social life.

Staying at the frontline of tech has its cost, however: the soft subject must commit to constant upskilling and endless 'crunch.' This is the problematic 'complicity' evident to greater or lesser degrees for all of those who make images using software. Can a soft subject engage with novelty without compounding obsolescence? Can they create any stable cultural value within systems that constantly mutate in their endless quest for finer resolutions of simulation and control? The examples of contemporary computer graphics (CG) practice I've explored elsewhere in my video essays *Spectacle, Speculation, Spam* (2016) and *Goodbye Uncanny Valley* (2017) show that commercial and experimental artistic cultures respond to this natively 'new media' dilemma in many different ways. To reinvest images with stable value, contemporary artists often retreat from manual production or resort to codified digital abstractions, while commercial production cultures often invoke the imagined pre-technicity of analogue media like stop-frame and cel animation.

I would suggest that jumping off the speeding train, urging it forward at ever greater speeds or pretending it doesn't exist are not the only options: soft subjects need not choose between regression, acceleration, or agnosticism. Reflexive practice begins with acknowledging the ontological and epistemological conditions of computational capitalism and being mindful of how it creates a relationship between labour, control, automation, crunch and obsolescence. If the media avant-garde is quickly the media-archaeological, then the soft subject should enter any engagement with image production technology knowing this, building in a conscious conceptual framework for their work as already-buried 'trace' of always-obsolete sociotechnical activity – indeed, this might become the overarching value of the work of art in an age of 'mechanical simulation.' As uncomfortable as this might be, it at least confronts the pain so many working in media production have: the sense that weeks, months and years of labour were devoted to creating aesthetic artefacts that tomorrow's audience would consider rudimentary, crude, low-res or easily achievable with shortcuts, presets or machine learning algorithms.

With this reflexivity as a principle of making images using software, obsolescence becomes a conscious precondition of creative engagement with tech, something essential whose value should be ready to salvage. What I'm proposing is that media practice as *the production of aesthetics* might also be considered equally as an *aesthetics of production* in which sociotechnical formations of labour are reflexively foregrounded, deliberated and valued. What this amounts to is not standing outside technology and pointing to its wonder or terror, but acknowledging that through endless variations on 'crunch,' human labour and machinic assemblages are compressed into social treasure, a commons that belongs to all. This shift in the framing of visual media production involves more than just establishing a creative ethic of 'artisanal' tech work, a diligent curatorial practice or a careful attendance to media archaeology: it means treating every media assemblage as a distinct instance of contingent technical evolution. McKenzie Wark puts this well when she says '[e]very hacker is at one and the same time producer and product of the hack, and emerges as a singularity that is the memory of the hack as process' (2004, p. 81). Marking the history of the soft subject's 'hacks' might one day lead to an account of how the proprietary formats, aesthetics and platforms of the 'metaverse' were sewn together by hand, by millions around the world in offices, mines, studios, bedrooms and factories.

Notes

1 These technologies involve the creation of synthetic 'worlds' that increasingly integrate with the real world. In my own recent video essay and exhibition, *RGBFAQ* (Warburton, 2020), I use my experience working in industry to explore how the twin fields of computer vision and computer graphics converge in recent GPU, software and smartphone technology to produce a pervasive and proprietary image regime that subjects the real world to the operational logics of simulation.

2 Similar developments can be seen in the evolution of rotoscoping (the 'low-level' cutting out of foreground from background) and camera tracking (where virtual 3D co-ordinates and camera positions are reverse-engineered from analysis of live action footage). Both are traditionally cumbersome and repetitive processes that rely on a combination of user skill and 'low-level' automation to get clean results. Both are expensive elements of the visual effects production line often outsourced from western VFX facilities to those in South-East and East Asia. Now, these aspects of production have been the focus of major 'innovation' through automated AI-assisted noise removal and feature detection baked in as standard to software like Autodesk Flame.

3 It's extremely difficult for researchers or even software developers themselves to audit software for complexity, but of the anecdotal accounts available it would appear that when version 1.0 of Adobe Photoshop was released in 1990, it comprised 100,000 lines of code, and by 2013 that number had grown to between 4.5 million (Shustek 2013) and 10 million (Paul 2013). Similarly, it is claimed that AutoCAD had a 'few thousand' lines of code when released in 1982 and in 2015 comprised 12 million (Hurley 2015).

4 Crunch might even just be one aspect of a longer and larger history of the relationship between computational media and exhaustion. Brod (1984) introduced the notion of 'technostress' in the workplace, Wurman (1989) described 'information anxiety,' Byung-Chul Han (2015) described the general exhaustion of a 24/7 mediasphere, and Kwon et al. (2020) catalogued the 'social media fatigue' felt by consumers worldwide. Digital creatives have often been at the centre of this story: in the late 1960s and early 1970s, artists like John Whitney and Larry Cuba would work nights at NASA's Jet Propulsion Laboratory (Carlson 2017), painstakingly producing labour intensive short films on military-grade mainframe computers. Similarly, accounts of work in early 1980s CG commercial studios are marked by tales of late nights, unpaid workers, burnouts, bankruptcies and suicide (Sito 2013).

Bibliography

Adler, Paul S. 1990. "Marx, Machines, and Skill." *Technology and Culture* 31 (4): 780.

Arthur, W. Brian. "Competing Technologies, Increasing Returns, and Lock-in by Historical Events." *The Economic Journal* 99 (394): 116–131.

Brod, Craig. 1984. *Technostress: The Human Cost of the Computer Revolution*. Reading, MA: Addison-Wesley.

Canizares, Galo. 2019. "Everything Is Software." *BLACK BOX: Articulating Architecture's Core in the Post-Digital Era*. Pittsburgh, PA: ACSA Press.

Carlson, Wayne E. 2017. *Computer Graphics and Computer Animation: A Retrospective Overview.* Columbus, OH: Ohio State University. https://ohiostate.pressbooks.pub/graphicshistory.

Chun, Wendy H.K. 2016. *Updating to Remain the Same: Habitual New Media.* Cambridge, MA: MIT Press.

Cote, Amanda C., and Brandon C. Harris. 2021. "The Cruel Optimism of 'Good Crunch': How Game Industry Discourses Perpetuate Unsustainable Labor Practices." *New Media & Society* (May 2021). https://doi.org/10.1177/14614448211014213

Davis, Jenny L., and James B. Chouinard. 2016. "Theorizing Affordances: From Request to Refuse." *Bulletin of Science, Technology & Society* 36 (4): 241–248.

Deuze, Mark and Mirjam Prenger. 2019. *Making Media: Production, Practices, and Professions.* Amsterdam: Amsterdam University Press.

Dewdney, Andrew, and Peter Ride. 2014. *The Digital Media Handbook.* 2nd ed. London: Routledge.

Fuller, Matthew. 2003. *Behind the Blip: Essays on the Culture of Software.* Brooklyn, NY: Autonomedia.

Fuller, Matthew. 2005. *Media Ecologies: Materialist Energies in Art and Technoculture.* Cambridge, MA: Leonardo, MIT Press.

Fuller, Matthew. 2006. *Softness: Interrogability; General Intellect; Art Methodologies in Software.* Aarhus: Center for Digital Æstetik-forskning.

Gaboury, Jacob. 2021. *Image Objects: An Archaeology of Computer Graphics.* Cambridge, MA: MIT Press.

Galloway, Alexander R. 2012. *The Interface Effect.* Malden, MA: Polity.

Gehl, Robert W., and Sarah A. Bell. 2012. "Heterogeneous Software Engineering: Garmisch 1968, Microsoft Vista, and a Methodology for Software Studies." *Computational Culture* 2. http://computationalculture.net/heterogeneous-software-engineering-garmisch-1968-microsoft-vista-and-a-methodology-for-software-studies/

Geoghegan, Bernard D. 2019. "An Ecology of Operations: Vigilance, Radar, and the Birth of the Computer Screen." *Representations* 147 (1) 1: 59–95.

Han, Byung-Chul. 2015. *The Burnout Society.* Stanford, CA: Stanford University Press.

Hesmondhalgh, David, and Sarah Baker. 2008. "Creative Work and Emotional Labour in the Television Industry." *Theory, Culture & Society* 25 (7–8): 97–118.

Hoelzl, Ingrid, and Rémi Marie. 2018. "On the Future Evolution of the Image." In *The Evolution of the Image*, edited by Marco Bohr and Basia Sliwinska. New York/London: Routledge, 134–147

Hurley, Shaan. 2015. "Throwback Thursday – 12 Million Lines of Code." *Between The Lines.* 9 April 2015. https://autodesk.blogs.com/between_the_lines/2015/04/throwback-thursday-12-million-lines-of-code.html.

Kwon, Eunseon P., Ashley E. English, and Laura F. Bright. 2020. "Social Media Never Sleeps: Antecedents and Consequences of Social Media Fatigue Among Content Creators." *Journal of Social Media in Society* 9 (2): 150–172.

Lorusso, Silvio. 2017. "What Design Can't Do – Graphic Design between Automation, Relativism, Élite And Cognitariat." *Institute of Network Cultures.* https://networkcultures.org/entreprecariat/what-design-cant-do/.

Mackenzie, Adrian. 2006. *Cutting Code: Software and Sociality.* New York, NY: Peter Lang.

Mackenzie, Adrian. 2017. *Machine Learners: Archaeology of a Data Practice.* Cambridge, MA: MIT Press.

Manovich, Lev. 2002. *The Language of New Media.* Cambridge, MA: MIT Press.

Manovich, Lev. 2006. "Image Future." *Animation: An Interdisciplinary Journal*, 1 (1): 25–44.

Manovich, Lev. 2013. *Software Takes Command.* New York; London: Bloomsbury.

Paul, Ian. 2013. "Computer History Museum Shares Original Photoshop Code." *PCWorld*. www.pcworld.com/article/456862/computer-history-museum-shares-original-photoshop-code.html.

Peckham, Eric. 2019. "How Unity Built the World's Most Popular Game Engine." *TechCrunch*. https://techcrunch.com/2019/10/17/how-unity-built-the-worlds-most-popular-game-engine/.

Richardson, Lizzie, and David Bissell. 2019. "Geographies of Digital Skill." *Geoforum* 99: 278–286.

Shustek, Leonard. 2013. "Adobe Photoshop Source Code." *Computer History Museum*. https://computerhistory.org/blog/adobe-photoshop-source-code/.

Sito, Tom. 2013. *Moving Innovation: A History of Computer Animation*. Cambridge, MA: MIT Press.

Tom, Edith, Aybüke Aurum, and Richard Vidgen. 2013. "An Exploration of Technical Debt." *Journal of Systems and Software* 86 (6): 1498–1516.

Warburton, Alan. 2016. Spectacle, Speculation, Spam. Video. https://alanwarburton.co.uk/spectacle-speculation-spam.

Warburton, Alan. 2017. Goodbye Uncanny Valley. Video. https://alanwarburton.co.uk/goodbye-uncanny-valley

Warburton, Alan. 2020. RGBFAQ. Video. https://alanwarburton.co.uk/rgbfaq.

Wark, McKenzie. 2004. *A Hacker Manifesto*. Cambridge, MA: Harvard University Press.

Wark, McKenzie. 2019. *Capital Is Dead: Is This Something Worse*. London: Verso.

Wurman, Richard S. 1989. *Information Anxiety*. New York: Doubleday.

PART III

Curating the Networked Image

7

THE PARADOXES OF CURATING THE NETWORKED IMAGE

Aesthetic Currents, Flows and Flaws

Gaia Tedone

Introduction: Curating in an Overtly and Overly Curated Environment

The networked image is a slippery image as much as it is a slippery concept: it not only resists being framed, in the material sense of the term, but also it eludes a fixed definition. Yet, it is precisely its slipperiness that can open up multiple access points into its conceptualisation and curation. Through the pages of this book and in the scholarship of its contributors, the networked image has taken up a flexible role, coming to account for the image as 'a dynamic, distributed and computational object that unsettles received notions of space-time' and for network as 'a descriptor of dynamic social relations as much as technological infrastructure' (www.centreforthestudyof.net). Although in this chapter I maintain an elastic definition of the networked image, I also retain the usefulness of the concept to describe more specifically the dynamics of online image searching and the circulation of images within computational culture and I look at the impact of these dynamics on the practice of curating.

Currently, within the context of networked culture, the position of the curatorial function is similarly elastic, as disparate professional fields and discourses coexist under the same semantic rubric. In other words, online a new definition of curation is only ever one click away. While the literature from the field of curatorial studies (Levi Strauss 2006; O'Neill and Wilson 2014; Smith 2015) continues to validate the role of the curator as a cultural operator and as a 'free agent, capable of almost anything' (Levi Strauss 2006) within the domain of art, the work of creative entrepreneurs and business scholars (Dale 2014; Rosenbaum 2011) posit curation as the savvy use of data aggregation tools and real-time technology to accrue competitive advantage within the online marketplace.

DOI: 10.4324/9781003095019-11

When looking more specifically at the context of online popular culture, the procedures of curating interlock with the operations of the networked image since both are increasingly defined by the political impact of algorithms in culture and society (Dewdney 2019) and the flows of visual data across networks. Online the activity of curating has become heterogeneous, abundant and massively distributed (Tyżlik-Carver 2017a, 2017b) and every user can potentially engage with some form of curating through the organisation, filtering and arrangement of digital images and information. These forms of curating, which are commonly referred to as content curation, digital curation or, more broadly, as online curation, are shaping the professional field of 'the 21st century knowledge worker' (Dale 2014) and are supported by an emerging literature and a range of digital data aggregation tools and software.[1]

Online curation crucially relies upon the work of algorithms and software for searching, collating, sorting and visualising networked images, which are read as 'data' and 'visual information to be analysed and remapped to new contexts by algorithms' (Rubinstein and Sluis 2008, p. 21). Additionally, it is embedded in the everyday use of social media platforms and networked devices. As such, it is both openly executed by digital professionals, online users and software, and excessively performed, since it is virtually applied to all sorts of commercial contexts and situations – any sellable product or service can now be virtually curated.[2]

It is around this current curatorial exuberance that distinct and at times rival curatorial paradigms and definitions coalesce: art curating, new media curating and content curation. Rather than dismissing such hybridisation, this chapter considers it the ground upon which the curation of networked images fundamentally takes place and posits circulation as the condition which enables it. The two core examples discussed in this chapter, one of which is drawn from my own curatorial practice, inhabit precisely this territory and make visible a number of paradoxes concerning the online curation of networked images. While the first example revolves around an investigation of a particular image, an H&M T-shirt patterned with the statement 'this image is not available in your country', the second example focuses on collecting and archiving a stream of images – predominantly memes – posted on a dedicated Facebook page by artists Félix Magal and Emilie Gervais under the umbrella of their collaborative web project 'MoI: Museum of Internet' (2012–2019). Through their analysis, these two case studies illuminate the challenges and possibilities that exist when the object of curating is no longer a conventional art object but a poor, authorless, bastard image, and when curating steps outside the 'art filter bubble' and turns into networked co-curation[3]: a process forging strategic alliances between different agents (both humans and machinic), objects (art and mundane objects) and practices (independent and institutional curating, hacking and commercial endeavours), and that aggregates each time a specific constellation of relationships (Dekker and Tedone 2019; Tedone 2019).

To set the framework for the analysis of the case studies, I will provide a theoretical exploration of the concept of circulation to highlight how the latter functions as the link between the operations of the networked image and the function of online curation. In what follows, I want to propose that an integrated perspective on circulation serves the purpose of framing the politics and aesthetics of curating the networked image.

The Politics and Aesthetics of the Networked Image: Framing Circulation

When discussing and experimenting with what it means to curate the networked image, circulation stands as a central concept which illuminates both a key operational behaviour of the networked image and an atmospheric condition that affects the various affordances of curating within networked culture, as mentioned above. On the one hand, circulation is posited as one of the defining and operational conditions of the networked image, enabling its visual surface and computational structure to interact with various contexts of reception and users. On the other hand, circulation is recognised as having important implications on conditions of online curation, pointing to problems of commodification, aesthetics and interpretation, as the discussion of the case studies will reveal.

The rapidity with which images circulate across the Internet has multiplied their contexts of reception and patterns of interpretation. This poses some challenges for the work of the online curator who neither relies on the physical space of the art gallery nor on the photographic frame to designate context and fix meaning. Through rapid processes of circulation, reproduction and self-replication, the networked image reaches a variety of contexts of reception and users and gets embedded within a specific socio-technical network, which in turn partakes in the creation of its meaning, value and function. Online circulation thus becomes one of its defining and operational conditions, as it enables its visual surface and computational structure to interact with various environments and users. Whilst scholars from diverse fields, including media theory and art history, have been emphasising the relevance of the concept of circulation in relation to the currency and value of digital images (Joselit 2013; Olson 2015; Steyerl 2012), little attention has been paid to the implications of circulation for curatorial practice. Here I want to argue that as networked images can no longer be fixed nor 'framed', in the conventional sense of the term, curators need to be equipped with new conceptual tools to engage with their ubiquity (Hand 2012) and undecidability (Rubinstein and Sluis 2013a, 2013b).

When the domain of the networked image and that of online curating are brought together and jointly analysed, circulation emerges as the key socio-technical process that enables the distribution and sharing of online content (objects, images, data) across various users and contexts of reception. In addition, a broader look at this concept that considers its uses also in different fields, will serve to further illuminate its wider political and aesthetic implications with respect to the curation of networked images. It has to be noted that circulation is a complex concept that is widely used in science and technology studies as well as in media and cultural studies, but discussions of it have nevertheless remained largely generic. A dedicated issue of the *Italian Journal of Science & Technology Studies* (Balbi, Delfanti and Magaudda 2016) has attempted to address this problem, offering an integrated perspective of circulation that foregrounds the need to bring together considerations about media, materiality and technological infrastructures. This integrated approach to the notion of circulation – or this 'critical perspective on circulation' (Lee and LiPuma 2002, p. 192) – is considered essential to approach circulation as an open-ended cultural process, which cuts across different socio-technical and economic layers and greatly impacts dynamics of sharing and cultural appropriation as well as systems of exchange and commodification. In other words, this integrated perspective on circulation offers a framing of the politics and aesthetics of the networked image. This is because it considers how networked images are simultaneously part of the wider networked infrastructures and dynamics of global capitalism and of the cultural flows and aesthetic currents that define what it means to live, create, consume and imagine in the age of computing. The concept and metaphor of the flow have been used widely in the literature for describing the paradigmatic shift of the digital image from a singularity to a multiplicity (Henning 2018; Joselit 2013; Jurgenson 2019) as well as in discussions about the unstable condition of art on the Internet (Groys 2016). Here I employ terms such as currents, streams and flows, to account for an aesthetics which originates from specific socio-technical assemblages, such as art platforms (Goriunova 2012), and mobilises the creative inputs of networked users and technical agents.

When streaming through these aesthetic flows and currents, networked images operate as socio-cultural entities that spin out of the control of their makers, through dynamics of reproduction and acceleration, which are executed by software, automatised procedures and algorithmic feedback loops. As a result of these circulatory movements, they entangle themselves in processes of commodification and become subjected to the logic of cultural appropriation, a deep current which challenges traditionally accepted notions of artistic originality and single authorship. Artist and theorist Hito Steyerl is a key commentator of circulation as the social process of image sharing and distribution. She describes it with her own term – 'circulationism' – which is the 'art of postproducing, launching, and accelerating an image rather than producing it' (Steyerl 2013). She contends that on the web, production and circulation are mixed up to the point of being indistinguishable and points to the need to be critical of the very patterns that image circulation produces across networks.

As Steyerl (2013) observes:

> Circulationism, if reinvented, could also be about short-circuiting existing networks, circumventing and bypassing corporate friendship and hardware monopolies. It could become the art of recoding or rewiring the system by exposing state scopophilia, capital compliance, and wholesale surveillance.

In this broad sense, the concept of 'circulationism' can be employed to formulate a more creative and political approach to the wider dynamics of image circulation, inviting the online curator to consider not only 'the circulation of forms' – which networked images circulate and how – but also crucially 'the forms of circulation' (Appadurai 2010) – how circulation itself operates, which power structures it relies upon or reinforces and what kind of invisible processes it activates or obstructs. It is here maintained that the discontinuous and performative nature of circulation – one which binds the distribution of content to specific geopolitical and infrastructural conditions, governmental policies and corporate strategies – affects the conditions under which economic and social value is produced within networked culture and calls for a certain operational agility on the side of the online curator.

The case studies discussed below intend to explore how this operational agility is manifested when curating both in the wild, by which I mean here the larger context of the Web 2.0 and Web 3.0,[4] as well as in the context of a specific social media platform, in this case Facebook. While the first case study will bring us all the way to Cuba to show that networked images are part of a 'very material infrastructure, and their circulation is regulated, slowed down or blocked according to the same boundaries and borders we encounter in our own movements' (De Mutiis 2020), the second case study will examine what happens when their flow is re-channelled and archived on a commercial platform and under the agenda of a fictitious, yet progressive, museological operation. As such, it will open up a more general reflection on the fraught relationship between museums and networked culture – a relationship whose complexity escalated as a result of the Covid-19 pandemic and the forced migration online of many cultural institutions.

Flows and Blockages: This Image Is Not Available in Your Country

In January 2016, I purchased from a charity shop in South London a second-hand H&M T-shirt (Figure 7.1) that had printed on it the phrase: 'this image is not available in your country'. This phrase, which is usually associated with the removal of content from the web for censorship and copyright reasons, opened up some thought-provoking questions on the conditions of availability and the circulation of images within the networked culture: where and what was the networked image (actual or potential) that was supposedly claimed by this T-shirt? Was the printed phrase a kind of networked image in itself? Or was the networked image the resulting selfies that were anticipated by H&M's visually savvy branding strategy? Or, again, was the networked image to be found at the meta-level prompted by its message, that is, in the networks of invisibility and paths of censorship of an allegedly interconnected global world?

In order to answer these questions, I pursued two tracks of investigation, which encompassed tracing the circulatory patterns of the T-shirt as a material object as well as a networked-screen image.[5]

My inquiry into H&M's distribution chain revealed that the T-shirt only featured on the US website where it was sold out, and it was not possible to track whether it

FIGURE 7.1 This Image is Not Available in Your Country, H&M T-shirt, Bangladesh. Image by Gaia Tedone

was purchasable from any other H&M website or store. Hence, the item of clothing was, in a very real sense, 'unavailable in my country' as well as everywhere else. In fact, even if the T-shirt had been available on the US website, I would not have been able to purchase it, as one cannot purchase an item online unless one is able to pay in the local currency. These findings revealed how within the current state of the commercial web, the online environment mimics a networked economy that gives the illusion of smooth global transactions but is actually deeply permeated by economic and geographic boundaries.

The second track of investigation I pursued was that of exploring the T-shirt's relation to screen culture and the patterns of circulation of this specific networked image across the Internet. The photograph of the H&M T-shirt was widely accessible online: it could be found on personal fashion blogs, search engines results, social media and 'idiotic' (Goriunova 2013) imageboards, including, for instance, 'themetapicture.com'. The T-shirt was photographed, posted, shared and commented on by Internet users (socio-technical circulation) and its image used by e-commerce sites, such as the Los Angeles-based commercial platform 'Society 6', for the creation and distribution of other items of merchandise (economic and commercial circulation). These included blankets, iPhone covers, fine art prints and shopping bags. The printed statement inspired various artists in the titles given to their works and exhibitions (cultural circulation).[6] Throughout this investigation and curatorial experiment, the recirculation of this particular networked image was further accelerated. For example, I deliberately shared the image across multiple platforms, channels and formats, including conferences, publications, online workshops, as well as hard drives and memory sticks.[7] In some cases this resulted in the image being recirculated outside of my control, evidencing the web's idiosyncratic logic of appropriation, and the discontinuous and performative nature of circulation. Visual content now transits to unexpected corners of the web as a result of algorithmic systems of recommendation and unfettered distribution of data by corporations. This occurred as a consequence of my visit to the H&M's main store in London where I located a vintage looking Photoautomat, in which I took a number of self-portraits wearing my H&M T-shirt. This gesture was intended as an act of reflexive self-exposure that would allow me to engage in a critique of the pervasiveness of mechanisms of 'commercial image-making, brand development or trend-setting' (Groys 2016, p. 129) associated with networked culture. This critique was performed through the embodiment of such mechanisms, whereby embodiment here denotes not only physical reality and everyday experience but also participatory action (Dourish 2001). Upon printing, the photostrips appeared with the H&M logo printed onto them, transforming my own image into a by-product of the H&M brand. When I shared the images on my social media accounts, they were immediately re-tweeted by some unknown user, or perhaps an Internet bot, related to H&M. In this sense, in my attempt at curating this networked image, my own image ended up being curated by H&M, fuelling a potentially endless vortex of commodification.

My examination of H&M's retailing system thus offered the opportunity to shed light on the entanglements between economic and political forces that mould the circulation of goods and images online and to recognise in those entanglements disguised forms of (cultural) regulation and control that the online curator needs to be critical of. For instance, H&M exploits the economic dependency of a less advanced economy, Bangladesh, to secure the production of goods in offshore factories at cheap labour costs. When one considers that the H&M T-shirt was in fact produced in one such factory, the meaning of the statement 'this image is not available in your country' – which invokes wider dynamics of inclusion and exclusion in global information capitalism, the so-called digital divide (Castells 2001) – becomes darkly ironic, if not perverse.

A closer analysis of the T-shirt's statement also offered the opportunity to reflect upon circulation as a process that is entrenched with a proliferation of invisible boundaries, ruptures and enclosures. As previously observed, this sentence is commonly associated with the impossibility of accessing content on the web, particularly on online platforms such as YouTube, because of specific geo-localisation restrictions prompted by issues of copyright and censorship or ascribed to the practice of geo-blocking. The latter refers to the use of technology to restrict access to Internet content, based upon the user's geographical location. Such a practice is employed by online platforms and e-commerce sites, such as H&M, which filter international audiences and customise their offerings according to territory, language and advertising markets. As media scholar Ramon Lobato observes:

> Like search localisation and algorithmic recommendation, geoblocking is a 'soft' form of cultural regulation. Its widespread adoption is changing the nature of the open internet by locating users within national cyberspaces and customising content based on certain ideas about territorial markets.
>
> *Lobato 2016, p. 10*

The 'softness' Lobato refers to is itself geographically and politically dependent, as it varies according to different types of networked infrastructures, speeds of access to the Internet and uneven conditions of connectivity. This aspect became particularly tangible in my subsequent presentation of this project as part of the online programme of 'Screen Walks' (2020) – an experimental curatorial format conceived by the digital curators of The Photographers' Gallery in London and Fotomuseum in Winterthur in response to the Covid-19 pandemic. On that occasion, I invited artist and collaborator Nestor Siré, who is based in Cuba, to recount our joint enterprise of circulating elements of my curatorial project around the H&M T-shirt/image inside of the *Cuban Paquete Semanal*. This is a one-terabyte collection of media that is aggregated weekly in Cuba on a physical hard drive and is distributed through a pirate network of distribution via in-person digital copying. Inside the *Cuban Paquete Semanal* Siré curates the 'Art Section' – a folder which mimics in ethos and organisation the overall structure and curatorial intelligence of *El Paquete*, but is specifically dedicated to spreading information, news and content about art to the local community of artists and creative practitioners. The art folder, which is released on a monthly basis, thus serves as an offline repository of curated content, but also functions as a portable art gallery or exhibition showcasing site-specific interventions and projects that reason with the particular format and logic of *El Paquete*.[8]

Since the 'Screen Walk' took place on Zoom – a platform which at that time was not accessible from Cuba – and the format allowed for performative experimentation,[9] the joint presentation was approached as an opportunity to give visibility to a number of blockages Cuban users experience in their daily navigation of the Internet. These blocks concern both websites which are banned 'internally' by the Cuban Government for political reasons, including journalist blogs, as well as commercial services that are blocked 'externally' because of the embargo with the United States, such as PayPal, Adobe and the app for the H&M store. As Siré describes in a video contribution[10] produced ahead of the 'Screen Walk' and which was live streamed during the event, 'Cubans are blocked on both sides' (Siré 2020).

The opportunity to perform this collaborative presentation under conditions of 'digital asymmetry'[11] was illuminating with regard to the specificity of the Cuban case and, more broadly, for apprehending how conditions of digital divide are not just political, social and technical but also cultural and generational. In a concluding remark to his video, Siré states that his mum thinks that 'the Internet is Facebook' (Siré 2020). Here, again, the irony of this statement is quite dark as it perceptively points to the conditions of commercial and cultural monopoly that American corporations exert online. The question of what it means to curate under these conditions is one that artists and curators working online have been tackling for a while. Their responses, often, take the form of hacking strategies and tactics of infiltration that subvert the very mechanics of online platforms.[12] In the next section, I will look more closely at one of such projects which took place on Facebook and exposed a different kind of digital divide, one that I refer to as the 'curatorial digital divide'.[13] This divide pertains to the way in which digital art, and more generally networked culture, has been dealt with by the mainstream contemporary art system, its institutions and curatorial discourse. So, it is a divide of fields, institutions of reference and intellectual discourses.

Aesthetic Currents and Flaws: The Museum of Internet (MoI) of Félix Magal and Emilie Gervais

The Museum of Internet (MoI) was an online project created by artists Félix Magal and Emilie Gervais which was hosted on Facebook from 2012 to 2019. Responding to the lack of interest in Internet culture by the Parisian art scene and its institutions, Magal and Gervais built their own museum in the form of a Facebook page *on* the Internet and *about* the Internet, archiving the flow of content representative at that time of Web 2.0 culture. Outside Facebook's interface, the museum existed simply as a folder, containing what Magal (2019) describes as 'a lot, lot, lot of memes'. Inside of Facebook, the museum was a content feed of images posted and streamed on a daily basis, which captured the daily temperature of the Internet, its aesthetic currents and the online iconography emerging around the 2010s. The history of MoI was bound to the socio-technical specificity of its hosting platform of its hosting platform, which was considered at the time by its founders as the most apt one for this type of project because of its traffic and population. In other words, unlike today's scenario, Facebook used to be the platform where 'the new kinds and manners of aesthetics' (Goriunova 2015) were gaining traction: representation of everyday life on the Internet, testimonies of the absurd, photos taken on the spot and spontaneous creation.

The Facebook guidelines regulated and moderated the project's aesthetic currents and set the limits of MoI's collection – no nudes, no hate speech and no violence – performing a form of self-censorship and regulation akin to that of a government. To counter this algorithmic gatekeeping, Magal and Gervais also created a separate website where users were able to upload images without restrictions: there, unsurprisingly, the aesthetic current moved downstream towards the porn shore, forcing Magal and Gervais to eventually take up the role of content moderators. When the Facebook page closed in 2019, as the platform was shrinking in popularity and ethical concerns were overflooding its curatorial procedures,[14] the project's online presence survived in the form of an archival website organised by tags, which is currently in the process of being redesigned. Like most projects of this nature, its destiny and survival remain unclear to date.

As a museum, MoI was indeed quite atypical. It featured no exhibition and its founders refused curatorial authorship and stardom, as Magal and Gervais decided to remain anonymous. MoI provided no interpretation since memes were shared without content and captions. Captions, in Magal's view, are utterly unnecessary when it comes to curate memes. As he states: 'if you put a caption on a meme you have no reason to live' (Magal 2019). Underneath this provocation lies a hard truth: that 'the digital as culture challenges not only the existing organisation of the museum, but also its categorical distinctions' (Dewdney 2019). Instead, MoI chose to give prominence to its content and users, exercising a form of radical 'communality' (Stalder 2018) in order to democratise aesthetics. As a result, the operation demonstrated that the network is *already* in the museum by virtue of its audiences (Dewdney 2019). In other words, the project celebrated Internet art, or more precisely, it celebrated Internet *as* art, and thus art as a potentially accessible, participatory and humorous event and practice. Throughout the process, MoI grew a community of over one million followers spread across more than one hundred countries.

Yet, there is something particularly revealing about this project being called a 'museum'. For Magal, the title of the project played to a certain collective perception that he felt compelled to address for two contradictory reasons: on the one hand, in his view, kids who are interested in Internet culture tend to avoid going into museums; on the other, museums do not understand the Internet and are often sceptical to embrace online culture.[15] This can be considered somehow paradoxical, as the Internet is continuously producing a new aesthetics and there is no place where to archive it, whilst museums approach 'the potential of the web to produce a universal archive, cast as digital heritage' (Dewdney, Dibosa and Walsh 2013, p. 330). Gervais also speculated on this conundrum when she observes:

> Too bad museums are always late and getting on with stuff once it's history. I often wonder if it is that way because they want to make sure the general public is following them or because they need to have an historical point of view on things once they can grasp it as an actual thing.
>
> *Gervais 2019*

The project's denomination and its archival mandate suggests the urgency of narrowing the divide between the museum and networked culture and points to the necessity of reassessing the very premise of the museum's mandate, starting from its formats and operations. Under this light, what arguably presented itself as a discrete, almost accidental and light-touch artistic operation that resisted over-theorisation[16] provides today a valuable insight when assessing the complicated relationship between the museum and networked culture. Particularly after the year 2020 when, for mainstream cultural institutions, migrating online became not a choice but a forced condition.[17] It can be argued, in fact, that the Covid-19 pandemic has scaled up an earlier identity crisis of the museum, which has been struggling for a while to face the social, educational and aesthetic challenges brought about by networked culture and to explore its creative and aesthetic potential.

As museums began to craft OVR (Online Viewing Rooms) on corporate services such as Google Arts and Culture, treating them as if they were transparent interfaces, a great opportunity was missed: that of conducting a systematic operation of archiving the richness and variety of memes and visual content that was created and exchanged by users worldwide during the lockdowns via WhatsApp chains and social media platforms.[18]

I can only speculate that the MoI would have been up for the task, performing a cross-geographical comparison of meme culture at the time when humour was the only antidote to despair. As the aesthetic current of the global pandemic began to decelerate in early 2021, another strong participatory wave surfaced: the phenomenon of the Bernie Sanders meme and the virality of the site 'Bernie Sanders goes places'.[19]

This was a particularly interesting case not only for its spread and the kind of performativity it produced – a truly networked performance with a wild geotagging flavour – but also from the resilience shown by the public figure whose image was being curated. Sanders benevolently accepted the spotlight and reclaimed his own image only for the greater good: to raise $1.5 million in support of Vermont Charities through the sale of his line of meme merchandise. How come, then, that the museum remains largely indifferent to these aesthetic currents and their wider social and political implications?

The answer to this question considers two distinct yet interrelated factors, both of which are a consequence of the unresolved clash between the old paradigm of 'aesthetic modernism' (Dewdney, Dibosa and Walsh 2013) and the current conditions of online circulation and 'circulationism' (Steyerl 2013) which were described above: first, the difficulty for the museum to radically rethink the notion of aesthetics; second, the resistance to reimagine its categorical distinctions, as mentioned above, as well as its curatorial formats and the whereabouts of its operations.[20]

Over the last years, the problem of how to describe this new mode of aesthetics has preoccupied scholars and curators from different fields and theoretical perspectives. A number of new qualifiers were put ahead of the term aesthetics to denote it, including: 'web aesthetics' (Campanelli 2010), 'new aesthetics' (Bridle 2012) and 'network aesthetics' (Joselit 2013). More recently, the discussion of 'memestetica' (2020) offered by curator Valentina Tanni provided a welcome contribution to the debate. Tanni focuses her attention on the function of memes as specific types of networked images and advocates in favour of an art which infiltrates into non-institutional places by adopting unexpected and inappropriate modalities. This position resonates with the acknowledgement that aesthetics is simultaneously a 'major mode of operation for contemporary society' (Goriunova 2012, p. 94) and a practice in a constant process of becoming whose very constitution is being changed by networked technologies. What still requires further attention, however, is an account of the implications that this mode of aesthetics has for the role, function and practice of both the museum and the curator. Both of them are nodes of a much wider constellation or complex ecosystem, which is shaped by human and non-human agents, software, images and online users. In this ecosystem, the curation of images happens as a networked process performed by various agents and profoundly shaped by the technology it relies upon. The aesthetic flows it accelerates can virtually take the curatorial process anywhere: on T-shirts and merchandise lines; inside hard drive or memory sticks; into folders on computer desktops. As we have seen through the example of the 'Screen Walk', when technology is treated as culture and the logics of cooperation and techno-social alliances replace that of cultural gatekeeping and single authorship, it is pos-sible to stage synchronic participation and interaction with audiences connecting from distant geographies and under different socio-technical conditions.

Conclusion: The Paradoxes of Curating the Networked Image

The examples discussed above show that the curation of networked images brings about a paradigmatic shift to the practice of curating which concerns both its object (a slippery image), its context (the wild web or the socio-technical assem-blage of a commercial platform) and the agents partaking its process (users, algorithms, software). This paradigmatic shift, I want to argue, does not come without its own set of challenges and paradoxical conditions. Quite the opposite in fact: curating the networked image is deeply imbricated in paradoxes. Two key paradoxes have emerged most vividly from the analysis of the case studies: the first can be synthesised in the Hamlet-esque dilemma 'to curate or be curated?'; the second more broadly concerns the tension between an excess in/of visibility and invisibility.

The first case study has offered an insight into curating under conditions of computational capitalism, by showing how the operations of the curators can easily enter in direct competition or 'ontological conflict of differentiation' (Goriunova 2013, p. 29) with other forces among which capitalist ones. This is because the web, as a material infrastructure and online marketplace, is far from being a neutral arena, but is a space largely subdued to private and corporate interests that influ-ence the behaviour of images, objects and users alike. Such a state of affairs poses some urgent challenges to both the operations of curators and museums. Not only must they take into account the increasing commodification of the web when pro-ducing and circulating content online and when acting on social media; they are also called to circumvent, disentangle and expose the uneven power structures and corporate dynamics of the networks. In order to do that, it is necessary to acquire a 'new vocabulary of visual language on a mass scale' (Goriunova 2013, p. 30) and to be trained in 'seeing aesthetic differentiation within the muddy swathe of cul-tural stuff' (Goriunova 2013, p. 29). Aesthetic differentiation does not aim to dis-avow cultural hybridisation, but rather develops in response to a critical perspective on circulation. As such, it can be employed as 'a nuanced gradient' (Goriunova

2012, p. 13) for seeing, recognising and dissecting aesthetic value and meaning outside and before corporate appropriation and for articulating a form of resistance towards consumerist forces and capitalist behaviours. This form of resistance does not unconditionally amount to refusal or rejection – as swimming against the deep currents of appropriation can be both exhausting and hopeless – but operates through a reflexive command of such mechanisms and a strategic rewiring of their codes and purposes.

The second paradox emerges out of the tension between a hypercurated scenario which operates at the interface of the networked image and an invisible, at times 'dark' curating, operating at the level of the algorithms and infrastructures supporting social media platforms; between an overabundance and excessiveness of creative content produced by users and an institutional vacuum when it comes to museums curating networked culture, as the analysis of the second case study revealed. This point connects with a more general condition that characterises the state of artistic and curatorial interventions online. These either exist within predefined contours that link them back to specific systems of reference and fields – these being for instance the 'filter bubbles' of the worlds of contemporary art and new media – perpetuating old institutional separations and curatorial divides that the logic of the web attempts to disrupt. Or, they risk dissolving within the plethora of creative content produced online or disappearing entirely from the web.

However, disappearance and invisibility are inevitable conditions of digital technologies which 'drive towards their own invisibility, their infusion into all material and physiological things and spaces' (Dewdney 2019). To navigate the thresholds between visibility and invisibility is precisely the aesthetic and political remit of curating the networked image. As a recursive, circular and multiple practice, which reflexively operates from a position of immersion within the same environment it interrogates and acts out its opacities, curating the networked image is concerned with exploring the negative spaces that exist between images and, by doing so, with exposing the power structures supporting their networked circulation. Here lies the value of adopting an integrated perspective on circulation when curating the networked image: to be able to remain attentive not only to what is visible, or representable, but also to what has disappeared or no longer is available.

Under this light, curating the networked image can be described as a practice of translation between different regimes of visibility, registers of code (the human and the computational), language (the visual and the semantic) and value (the cultural, the commercial and the social). As we have seen, the translation between the human language of signs and symbols, and the computational code of numbers and data, is not met without frictions and contradictions at the interface of the networked image. Rather, it is punctuated by misunderstandings and blockages at the hand of individuals, institutions and machines. When these flaws are carefully scrutinised to develop a deeper awareness of networked infrastructures, algorithmic systems and their resonating impact on society, then curating the networked image can enrich the vocabulary of visual language at a mass and micro scale. By paying attention to the systemic as well as to the granular, it can open up brackets of intervention into the contingent, distributed messiness of contemporary networked culture, moulding its politics and aesthetics.

Notes

1 The latter include, for instance, software solutions for content curation and content marketing, such as *Curata, flockler* and *scoop.it* among others.

2 This is the argument put forward by writer, researcher and digital publisher Michael Bhaskar who has addressed the spread of curation in his book *Curation: The Power of Selection in a World of Excess* (2016). As he obverses in the introduction, curation is 'at work everywhere, from the art gallery to the data centre, from the supermarket to our favourite social networks' (Bhaskar 2016, p.6).

3 This is a concept that I developed during my PhD research which was further elaborated through my collaborative writing with Annet Dekker in occasion of our joint article 'Networked Co-Curation: An Exploration of the Socio-Technical Specificities of Online Curation' (2019).

4 Although here I use the term 'in the wild' in its more current sense, in a forthcoming article co-written with Katrina Sluis and Nicolas Malevé (Malevé, Sluis, Tedone, forthcoming), the concept is developed further to analyse the conditions of curating vast datasets of images and to map an emerging range of visual practices that positions the computer science as a curator.

5 A full account of this case study can be found in my article for the *Journal of Media Practice* (2017) and my PhD dissertation (2019).

6 This is the case of the performance by Alexandra Stanciu at The Photographers' Gallery, London (2016), and the solo show by Ghana-born, Chicago-based Solomon Adufah at Surplus Gallery, Carbondale, IL (2017).

7 Here I am referring to the circulation of the project inside the hard drive of the *Cuban Paquete Semanal*, which I will detail later, and to its presentation inside my PhD portfolio of practice – a memory stick, whose content was organised and curated according to the logic of circulation.

8 For a comprehensive account of the phenomenon of *El Paquete* and of its 'Art Section', please see the collaborative research and writing of artists Nestor Siré and Julia Weist featured on their websites and on the Rhizome Archive.

9 By using the device of 'screen sharing' as a mode of entry into an artistic and creative practice, the format enables artists and researchers to open the black box of their computers and blends the boundaries between a studio visit, a networked performance and an online workshop. See the website: https://screenwalks.com.

10 This was one of the strategies used to circumnavigate problems of connectivity during the event. Another was to connect with Siré via WhatsApp desktop application rather than Zoom for live interaction with the audience.

11 This is a condition which I am further exploring together with Nestor Siré and Nicolas Malevé within the framework of our collaborative research project funded by the Swiss National Science Foundation and developed within the post-photography research group at the Lucerne School of Art and Design.

12 See, for instance, the project *#exstrange* on eBay curated by Marialaura Ghidini and Rebekah Modrak: http://exstrange.com.

13 I talk more specifically about this concept in my PhD dissertation (2019) where I make a direct reference to Claire Bishop's notion of 'Digital Divide' (2012).

14 These concerns culminated in the Cambridge Analytica Scandal of 2018. However, a year before leaked policies regarding the platform's content moderation policies were already fuelling debate about the social media giant's ethics. See, for instance, the 2017 Guardian article *Revealed: Facebook's Internal Rulebook on Sex, Terrorism and Violence* by Nick Hopkins (Hopkins, 2017). For a more general account of the functioning of the Facebook algorithm, please refer to the work of Taina Bucher (2018).

15 This was Magal's experience, who had attempted to connect with the Parisian institutional scene priori to setting up the project. The situation changed once MoI became popular and Magal was approached to collaborate with prominent museums, such as the Louvre in Paris and Tate in London.

16 Both Magal and Gervais maintain a light-touch attitude when describing their project, putting the emphasis on aspects such as serendipity and agility when defining it.

17 I have talked in more detail about this during the symposium on online curating organised by Masaryk University. This was a two-part talk developed with curator Marialaura Ghidini entitled '2020 Digital Odyssey – # 1 The Inglorious Precedents; # 2 Online or Nothing' (Tedone, 2020).

18 Some research institutes attempted to do so. See, for instance, 'Humor during the global Corona Crisis', a study conducted by Giselinde Kuipers and Mark Boukes from KU Leuven and Amsterdam University. I also attempted this operation of archiving networked images created during the pandemic with students from the 2020 edition of the Master in Museology and Museography at Catholic University, Milan where I teach.

19 https://bernie-sits.herokuapp.com/

20 In addition to this point, it is worth noting how the lack of technical capacities also limits the conservation efforts of museums, as it will be explored in the following chapters of this volume.

References

Appadurai, Arjun. 2010. "Circulation~Forms." *The Salon* 2: 5–10. http://jwtc.org.za/the_sa lon/volume_2/arjun_appadurai_circulation_forms.htm.

Balbi, Gabriele, Alessandro Delfanti and Paolo Magaudda. 2016. "Digital Circulation: Media, Materiality, Infrastructures. An introduction," editorial to *TECNOSCIENZA: Italian Journal of Science & Technology Studies* 7 (1): 7–16.

Bhaskar, Michael. 2016. *Curation: The Power of Selection in a World of Excess*. London: Piaktus.

Bishop, Claire. 2012. "Digital Divide." *Artforum*, September, 434–432.

Bridle, James. 2012. "#sxaesthetic report." *Booktwo.org* (Blog) 15 March 2012. http://book two.org/notebook/sxaesthetic.

Bucher, Taina. 2018. *If … Then: Algorithmic Power and Politics*. Oxford: Oxford University Press Inc.

Campanelli, Vito. 2010. *Web Aesthetics: How Digital Media Affect Culture and Society*. Rotterdam: NAi Publishers.

Castells, Manuel. 2001. *The Internet Galaxy: Reflections on the Internet, Business, and Society*. Oxford: Oxford University Press.

Dale, Stephen. 2014. "Content Curation: The Future of Relevance." *Business Information Review* 3 (84): 199–205. https://doi.org/10.1177/0266382114564267.

Dekker, Annet, and Gaia Tedone. 2019. "Networked Co-Curation: An Exploration of the Socio-Technical Specificities of Online Curation." *Arts* 8 (3): 86. https://doi.org/10.3390/arts8030086.

De Mutiis, Marco. 2020. *Introduction to Screen Walk by Gaia Tedone* June 17. https://screenwalks.com.

Dewdney, Andrew. 2019. "The Networked Image: The Flight of Cultural Authority and the Multiple Times and Spaces of the Art Museum." In *The Routledge International Handbook in New Digital Practices in Galleries, Libraries Archives, Museums and Heritage Sites*, edited by Hannah Lewi, Wally Smith, Dirk vom Lehn, and Steven Cooke. New York: Routledge, pp 68–80.

Dewdney, Andrew, David Dibosa and Victoria Walsh. 2013. *Post Critical Museology: Theory and Practice in the Art Museum*. London: Routledge.

Dourish, Paul. 2001. *Where The Action Is: The Foundations of Embodied Interaction*. Cambridge, MA: The MIT Press.

Gervais, Emilie. 2019. Written Interview about the *Museum of Internet*.

Goriunova, Olga. 2012. *Art Platforms and Cultural Production on the Internet*. London: Routledge.

Goriunova, Olga. 2013. "Light Heavy Weight Curating." In *Speculative Scenarios: Or What Will Happen to Digital Art in the (Near) Future?*, edited by Annet Dekker. Eindhoven: Baltan Laboratories, pp 25–31.

Goriunova, Olga. 2015. "The Force of Digital Aesthetics. On Memes, Hacking, and Individuation." *The Nordic Journal of Aesthetics* 24 (47): 54–75. https://doi.org/10.7146/nja.v24i47.23055.

Groys, Boris. 2016. *In the Flow*. London: Verso Books.

Hand, Martin. 2012. *Ubiquitous Photography*. Cambridge: Polity.

Henning, Michele. 2018. "Image Flow: Photography on Tap." *Photographies* 11 (2): 133–148. https://doi.org/10.1080/17540763.2018.1445011.

Hopkins, Nick. 2017. "Facebook's Internal Rulebook on Sex, Terrorism and Violence." *The Guardian* May 21, 2017. www.theguardian.com/news/2017/may/21/revealed-facebook-internal-rulebook-sex-terrorism-violence.

Joselit, David. 2013. *After Art*. Princeton, NJ: Princeton University Press.

Jurgenson, Nathan. 2019. *The Social Photo: On Photography and Social Media*. London: Verso Books.

Lee, Benjamin, Edward LiPuma. 2002. "Cultures of Circulation: The Imaginations of Modernity." *Public Culture* 14 (1): 191–213.

Levi Strauss, David. 2006. "The Bias of the World: Curating after Szeemann & Hopps." *The Brooklyn Rail*, 8 December 2006. https://brooklynrail.org/2006/12/art/the-bias-of-the-world.

Lobato, Ramon. 2016. "Introduction: The New Video Geography." In *Geoblocking and Global Video Culture*, edited by Ramon Lobato, James Meese. Amsterdam: Institute of Network Cultures, pp 10–24.

Magal, Félix. 2019. "Presentation at the Event 'Curating Social Media.'" The Photographers Gallery, London, September 13.

Malevé, Nicolas, Katrina Sluis, and Gaia Tedone. forthcoming. "Curating in the Wild." In *Curating Superintelligences: Speculations on the Future of Curating, AI and Hybrid Realities*, edited by Joasia Krysa and Madgalena Tyżlik-Carver. DATA Browser 10. London: Open Humanities Press.

O'Neill, Paul, and Mick Wilson. 2014. "An Opening to Curatorial Enquiry: Introduction to Curating and Research". In *Curating Research*, edited by Paul O'Neill, and Mike Wilson. London: Open Editions, pp 11–23.

Olson, Marisa. 2015. "Lost Not Found: The Circulation of Images in Digital Visual Culture." In *Mass Effect: Art and the Internet in the Twenty-First Century Cambridge*, edited by Lauren Cornell, Ed Halter. Cambridge, MA: MIT Press, pp 159–166.

Rosenbaum, Steven. 2011. *Curation Nation: How to Win in a World Where Consumers Are Creators*. New York: McGraw-Hill Companies.

Rubinstein Daniel, and Katrina Sluis. 2008. "A Life More Photographic; Mapping the Networked Image." *Photographies* 1 (1): 9–28.

Rubinstein Daniel, and Katrina Sluis. 2013a. "The Digital Image in Photographic Culture: Algorithmic Photography and the Crisis of Representation." In *The Photographic Image in Digital Culture*, edited by Martin Lister. London: Routledge, pp 22–40.

Rubinstein, Daniel, and Katrina Sluis. 2013b. "Notes on the Margins of Metadata; Concerning the Undecidability of the Digital Image", *Photographies* 6 (1): 151–158.

Siré, Nestor. 2020. Video Contribution to 'Screen Walk by Gaia Tedone', June 17. https://screenwalks.com.

Smith, Terry. 2015. *Talking Contemporary Curating*. New York: Independent Curators International.

Stalder, Felix. 2018. *The Digital Condition*. Cambridge: Polity Press.

Steyerl, Hito. 2012. *The Wretched of the Screen*. Berlin: Sternberg Press.

Steyerl, Hito. 2013. "Too Much World: Is the Internet Dead?." *e-flux Journal*, (49). www.e-flux.com/journal/49/60004/too-much-world-is-the-internet-dead/.

Tanni, Valentina. 2020. *Memestetica*. Rome: Nero Editions.

Tedone, Gaia. 2017. "Tracing Networked Images: An Emerging Method for Online Curation." *Journal of Media Practice* 18 (1): 51–62.

Tedone, Gaia. 2019. *Curating the Networked Image: Circulation, Commodification, Computation*. PhD Diss., London South Bank University.

Tedone, Gaia. 2020. "2020 Digital Odyssey: Online or Nothing." online presentation at the Symposium Online Curating organised by Masaryk University, April 23.

Tyżlik-Carver, Magdalena. 2017a. "Curator | Curating | the Curatorial | Not-Just-Art Curating: A Genealogy of Posthuman Curating." *Springerin* (1). https://www.springerin.at/en/2017/1/kuratorin-kuratieren-das-kuratorische-nicht-nur-kunst-kuratieren/.

Tyżlik-Carver, Magdalena. 2017b. "Posthuman Curating and Its Executions: The Case of Curating Content." In *Executing Practices*, edited by Helen Pritchard, Eric Snodgrass, Magda Tyżlik-Carver. New York: Autonomedia, pp 171–189.

8
INTERNET LIVENESS AND THE ART MUSEUM

Ioanna Zouli

Introduction

Since the beginning of the Covid-19 pandemic in 2020, the majority of cultural organisations have had to close down and renegotiate their programming to adjust to the new restricted conditions of living. With their physical spaces closed and the postponement of exhibitions and events, art museums had to deal with the financial consequences of freezing their activities and imagine alternative ways to present their work to their audiences. Under these unfamiliar conditions, they had to quickly adapt to online environments, such as social media, streaming and gaming platforms, and the institutional website as the only available spaces of operating. This was challenging for art museums' standard practices of exhibiting and public programming that usually require the presence of visitors in the flesh. In order to retain a connection with their audiences and show resilience in this moment of crisis, museums had to migrate a large part of their programming to online platforms. One of the reasons why this was challenging is that over the past two decades, art museums, especially large museum brands, are focused on offering *experiences* to their visitors (Hein 2000), which are in their majority inseparable from their physical spaces. And while digital experiences are often part of a museum visit, for instance in the form of immersive and interactive installations or prompts for social media engagement, museums have proven slow in embracing the digital as a culture that can be part of the history of art they represent as well as in understanding networked ecologies as something beyond communication platforms (Bishop 2012; Dewdney 2019; Dewdney, Dibosa, Walsh 2011, 2013; Walsh and Dewdney 2017; Zouli 2018). The forced conditions of the pandemic therefore became a testing ground for museums' capacity to take advantage of the networked environments and harness the public's interest in online spaces.

DOI: 10.4324/9781003095019-12

The main online practices that museums engaged with, especially during the first period of the pandemic, from March 2020 until the summer of the same year, ranged from podcasts, live broadcasts and online screenings to access to online collections and archives as well as virtual tours. At the same time, museums' social media activity increased as a way to directly engage with the networked audience: an audience that was under lockdown measures, having mainly screen-based outlets for news and entertainment while at home.[1] Cultural organisations faced the demanding task of having to claim their share of the audience attention online, at a moment when video conferencing platforms, live video game streaming and over-the-top (OTT) media services, such as Netflix, YouTube or Amazon Video, were thriving (Koeze and Popper 2020; Stephen 2021; Vlassis 2020). Although museums often use platforms like the above as part of their cultural programming – running a YouTube channel for video content, for example – or as part of their everyday work practices – such as speaking with remote collaborators and artists via Skype or Zoom – these are ultimately services that support the institutions' principal activities that take place offline. However, in the case of the worldwide pandemic museums and other cultural organisations had to perceive these actors already ingrained in the Internet, as important partners or competitors in art production and cultural entertainment online.

On reflection, museums' response to the pandemic's relocation of social life online was limited to platforms and technologies that the organisations already had the experience of using. In large part, they invested more in the publication, production and dissemination of content through their pre-existent channels than sampling new online activities and spaces; a choice which is typical for museums when they "cannot equal competitors that natively leverage digital networks," as the writer Mike Pepi (2014) points out. In today's accelerating technological landscape, when it comes to engaging with online spaces, museums remain fixed on the role of the connoisseur and producer of content. Video production and Internet publishing are established museum practices for twenty years now and are central for the framing and circulation of institutional knowledge online as well as for the extension of cultural authority beyond the physical spaces of museum buildings. Regardless of how thought-provoking or diverse the results of these practices might be, the logic behind their production reproduces an editorial and televisual model of transmission which is linear – from the source of knowledge to the audience – and hence limited comparing to the possibilities of online spaces. To a certain extent, museums' caution to become more distributed and dialogic in their online practices is based on the difficulty to let go of the historical and self-evident value of art objects as carriers of cultural knowledge and of their physical spaces as the platforms for art enjoyment and education. As Andrew Dewdney highlights, "the network challenges the notion that the creation of art is carried out necessarily by a singular vision intended for a singular viewer who is the sole locus of intended meaning" (Dewdney 2013, p. 108).

Another important factor for the museums' slow adoption of online art practices and recognition of the network's history, aesthetics and audience is that the formulation of strategic decisions, the everyday museum practices and the development of online cultures are asynchronous. Digital development has been on the art museums' agenda for at least ten years, while some organisations also publish dedicated digital strategy documents every two or three years based on their mission and marketing targets. However, the aspirations of digital innovation and expanded audience engagement that these documents propose often show a limited understanding of networked ecologies as well as of the role that art museums could have in them. There is also an inconsistency in the pace of interpreting the technological present by the museum compared to the speed at which networked technologies transform. Adding to that, the responsibility to implement these strategies usually falls on the organisations' marketing or media departments since digital media and the Internet are habitually conceived as tools for communication and transmission. This conception is accurate but it importantly disregards the network's dimension as a distinct ecology or as a space of multiple actors – human and nonhuman – where art can happen. One of the primary objectives of digital strategies remains to sustain the interest of online audiences in a way that they will eventually visit, or return to, the physical space of the museum; to a "real" destination where visitors can consume the branded art experience (Higgins 2015). It is perhaps not surprising that many art museums found themselves off-balance during the lockdowns of 2020: with their buildings closed, the art experience seemed unanchored.

The unexpected conditions of the pandemic and the necessary move of cultural programming to online spaces revealed the limitations of the model of the digital that museums pivot on. The complications that emerged during this period may not be surprising, given the precedents of art museums, of which I am about to give a detailed discussion. In order to understand how art museums incorporate, include or exclude networked culture and online audiences, one needs to take into consideration the politics and logic under which they operate. Politics and practices that have been developing for years and tend to remain attached to the role of institutions as containers of knowledge and as brands.

Tate Live

This chapter brings forward the BMW Tate Live: Performance Room programme as a case in point, to further elaborate on the art museums' association with the dynamic social relations of the network. The analysis of the programme that follows derives from my PhD fieldwork as an embedded researcher at Tate from 2011 to 2016. My work specifically focused on what happens when digital culture becomes a variable in the relationship of the museum with its audience and it addressed this question through a focus on Tate's production and use of video content (Zouli 2018). My embedded position in the museum allowed me to study the concepts of "digital" and "audiences" at the level of Tate's everyday practices. My ethnographic account of the networks of practice inside Tate focused on the Performance Room case study, as well as on the conceptualisations about the digital and audiences and how these affect the museum's cultural programming.

Performance Room is an important moment in the history of Tate as it was the first time that the contemporary art museum staged live performance art on YouTube, available only for the online audience. For this project Tate experimented with a format that was new for its curatorial and production teams: on the one hand, the live streaming of performance art, and on the other, the use of YouTube as the stage for the presentation of the performances instead of its usual institutional spaces in London, Liverpool or St. Ives.[2] The live participation of the audience in the online event was a crucial aspect of the programme as well as one of its biggest challenges. The development of the programme reflects on the one hand Tate's difficulty in embracing the distinctive participatory culture of the digital and on the other a strong tendency of the museum to incorporate digital practices mainly as part of its marketing logic. Despite Tate's aspirations, the Performance Room experiment proved to be too challenging for it to become a paradigm for Tate's online programming. Other strands of the BMW Tate Live programme, which included performance events and exhibitions at the spaces of Tate Modern, eventually overshadowed Performance Room. Today, it remains relatively unknown as a case study, possibly due to its questionable success; besides, ambivalence does not fit in the context of museum branding and sponsorship under which the organisation standards "must radiate the positive" (Stallabrass 2013, p. 155).

The significance of the Performance Room project is that it was an initiative that tested Tate's habitual understandings and perception of digital programming. It provides a window onto how a museum adapts and is open to unfamiliar practices, without the pressure of a crisis such as a global pandemic but under institutional constraints and contrasting dynamics in place. It is an important case study also because it shows how a national art institution, which aspires to be "digital by default" and is valorised by other UK museums as a front runner in the field, remains attached to a broadcasting logic as the main paradigm in its online practices. Museum practices cannot be seen as separate from the organisational politics and hierarchies that produce them and as such they also frame the institution's relation to its public. The discussion that follows reflects upon the dynamics that Performance Room brought into view and how these dynamics can shed light on museums' current digital frameworks and their hesitance to take flight into networked ecologies.

BMW Tate Live Performance Room: What Happens When a Performance Becomes an Image Embedded in Software on a Screen

Performance Room was first introduced in March 2012 and it centred on the live streaming of performance art online. It was a strand of a larger Tate initiative, the BMW Tate Live programme that started in 2011 as a four-year partnership between Tate and the vehicle company BMW Group. As it was described in the press release, BMW Tate Live was a series of events dedicated to "performance, interdisciplinary arts and curating digital space" (Tate 2011). Performance Room was the first project that developed out of this partnership. The project consisted of a series of commissioned, live performances, which took place in Tate Modern with no audience in the physical space of the museum and broadcasted on Tate's YouTube channel with viewers watching live on their screens from across the world. The participating artists' works covered a variety of live art practices from dance to conceptual art and music performances.

The invitation to the commissioned artists was twofold: in the first part of the event, the artists developed a performance piece designed simultaneously for presentation in a room of the Tate gallery and for the camera. The audience were invited to watch the broadcast of the piece live from Tate's YouTube channel. At the end of the performance, the gallery space transformed into a talk show format, with the artist(s) and the curator of the programme staging a short in-conversation about the performance and their work. The audience watching live from their devices were invited to leave their comments and questions about the performance either on the YouTube live comment board or on social media platforms using the hashtag "BMWTateLiveQ." These questions and contributions would feed live the Q&A session to give a sense of an open dialogue between the artist(s) and the audience.

The structure of the Performance Room programme has its roots in Tate's broadcasting capacity as a producer of video content. Since 2007 the museum has developed the TateShots programme, which is run by the Tate Media in-house production team and focuses on the production of short, documentary-type, films with Tate-related and art-related content. The films are hosted on both the institutional website as well as Tate's YouTube channel and they have been one of the main ways for the museum to provide "extended educational interpretation materials in digital format" (Maculan 2008, p. 113). By the time that Performance Room launched, TateShots was a well-established webcasting programme, so the museum had the expertise of shooting and editing high-quality video content as well as circulating it to its channels.

Adding to that, Performance Room surfaced at a moment that the institution was turning its attention to online spaces. In 2010, Tate published an online strategy in response to what was described as a flooded website and a need for a different approach regarding the institution's online spaces. The online strategy outlined a redesign of the Tate website in order to visually and contextually represent everything that Tate does: from exhibitions to conservation, to public programmes and fundraising. This was seen as a way for the museum to become "more porous (…) so it is clear who is speaking and where authority lies" (Stack 2010). In the same year, the museum also published a *Tate Social Media Communication Strategy,* which officially introduced social media at the core of Tate's operation online (Ringham 2011). The document highlighted Tate's intention to broaden its audience reach by extending the use of platforms such as Facebook, Twitter and YouTube, as well as to see the users of those platforms as communities of interest instead of just marketing targets.

The new website that Tate launched in March 2012 offered new ways of accessing Tate's collection online as well as its educational and video content. That year the museum reached 2 million followers on social media, which was considered a big success, while the new website was the most visited of the sector in the United Kingdom (Tate 2013, p. 34). These developments provided the ground for further reflection upon the museum's online practices and its correspondence to the technological moment and market. The lead actor in the development of Tate's digital vision was John Stack, the Head of Tate Online from 2007 to 2015. Stack was highly active in the community of museum and media practitioners, and during his direction of the department, Tate was considered a leading light in the UK contemporary art institutions that were envisioning a new digital landscape. It was also a period that major museums, realising the need to be in tune with digital developments especially regarding marketing and audiences, started appointing managers and curators for digital programming (Fei 2013).

Stack recognised the importance of third-party platforms like YouTube, Facebook and Twitter as well as the museum's responsibility to both expand its activities in these platforms and actively engage with the audiences that already use them. Another key idea that he introduced was that digital thinking should cut across the institution as a whole and not be the work of just Tate Media or Tate Online departments. These thoughts resulted in a new digital strategy that Stack presented in 2013 with the subtitle "Digital as a dimension of everything" (Stack 2013). Tate's 2013 digital strategy, a "holistic digital proposition" as it proclaimed to be, signalled a shift of focus towards the digital as a broader culture of operating. This shift did not differentiate the museum's established role as a content producer and publisher, which remained key for the distribution of Tate's content. In addition, the document suggested a broader scope of digital activities including multimedia tours, digital publications, digital fundraising, ecommerce, and digital ticketing services. The expansion of digital activities also required strengthening the digital skills of Tate staff and cross-departmental collaborations. The strategy forwarded a new "networked way of working" that would expand Tate's digital community through blogging, social media engagement and interactive learning activities.

This strategy expressed the museum's enthusiasm for the digital, which was seen as a way to do things differently and to broaden the engagement with the audience. As a Tate producer explained to me in March 2012, there were two key directions of the museum's regenerated attention to digital at the time: first, "a need for an understanding from different departments at Tate – for instance the current and future curators – of what digital media might be able to bring to a project in a physical space" and, second, the need to enhance Tate's production of content by orienting it "in the places where they [the audiences] expect to find it" (Zouli 2018, p. 120). Along similar lines, the digital strategist JiaJia Fei wrote an enthusiastic article, after the presentation of Tate's digital strategy at the 2013 Museums and the Web conference, propounding massive digital changes across art institutions that could break down the barriers inherited from previous generations and hierarchies (Fei 2013). She also mentioned *Connections*, an online video series by the Metropolitan Museum of Art (2011), as a successful example of a project that promoted the new digital aspirations for collaboration. The idea behind this video project was that museum staff from different departments, "from curators to security guards," took part in "non-academic" video interviews about their favourite works in the museum collection (Fei 2013). In both Tate's strategy and The Met example mentioned here, the production of high-quality and accessible content leads museum digital plans. At the same time, audiences have a fundamental role in museums' digital practices but they are also expected to be quite passive in the scheme of participation; they are seen, for example, more as viewers of content or consumers of services than active users in the platforms that museums present their work. While Tate conceives the digital as a multilateral way of presenting the "authentic voice of the institution" (Stack 2013), it appears less interested in the "cultural import" (Dewdney, Dibosa and Walsh 2013, p. 220) of the people that belong in the places where the museum stages this voice.

Performance Room was thus an ideal ground to evaluate Tate's new digital vision since the programme was based on a mixture of different practices and gave priority to cultural production for the online audience. The newly introduced aspects such as live streaming and audience participation represented a significant scaling up of Tate's video strategy. However, they also presented new challenges for curators and programming staff that had to translate their usual practices for the online space. Despite Tate's expertise in curating and exhibiting live art, its unfamiliarity with the Internet as a platform for the presentation of live performance art made the project demanding. Furthermore, the production and delivery of Performance Room required a temporary blending of practices and collaboration between Tate departments as well as with external partners; a blending that was not necessarily easy to achieve in the everyday museum settings. The mixture of practices necessary for the programme to run disturbed the usual departmental routines of Tate professionals that normally have a clear separation of professional activities in the organisational structure (Dewdney, Dibosa and Walsh 2013).

Another aspect of the programme that proved to be complex was the active role of the online audience that had a significant value from the beginning of the project. Despite the fact that the audience watching the event online was invited to participate in a discussion after the performance, Tate could not predict in advance how many people would engage in these discussions and what their level of engagement would be. This called into question the usual assumptions about how audiences interact with a piece of art or a live event in the physical space as well as the degree of agency that the museum enjoys in this framework. One of the core issues that emerged as the Performance Room progressed was that the YouTube audience, which constituted the main audience for the live streaming of the performance, was not necessarily the same audience that would regularly attend a live art event at the museum. Adding to that, the viewers' ability to comment on the performance in real time created an unsettledness regarding the reception and interpretation of the programme. Since the invitation to watch and comment was addressed to Tate followers as well as the general audience of the YouTube network, neither the type nor the content of the questions and comments could be predicted by the museum. It was obvious from the start that the Performance Room project team did not conceive YouTube as an independent cultural form with its own characteristics and audience or as a visual archive of moving images where different perspectives coexist (Richard 2008, p. 141).

The first performance of the series tested these dynamics and it was decisive for the structure of the rest of the programme. For the inaugural Performance Room, the choreographer Jérôme Bel presented a variation of his 1997 piece *Shirtology*, performed by his collaborator Frédéric Seguette. The piece was live-streamed on Tate's YouTube channel on 22 March 2012 (Tate 2012). Right after the end of the performance, the video cut to a close-up of Nancy Durrant, art critic for *The Times* who was the event host. Overall, the broadcast resembled a live television pro-gramme, particularly due to the presence of the external host who facilitated the discussion with cues, breaks and questions as well as due to the use of pre-recording footage about Bel's work and Tate's work with performance art.

Durrant hosted the discussion in a journalistic style, constructively integrating the audience's questions into the discussion, affirming their live presence. The audience's reactions to the event were both engaging and unexpected. Many of the questions and comments that arrived in real time during the broadcast were, according to Tate staff, "bizarre" and uncomplimentary to the event (Pringle et al. 2014). There were several comments from confused members of the audience who weren't sure what they have been watching, what the performer was doing and why. This confusion was expressed both kindly, with innocent questions, as well as with more rough comments of ridicule. The fact that Tate had advertised this first performance widely on YouTube, rather than targeting specific cultural channels, meant that a large number of the viewers did not necessarily have an art back-ground or were not used to viewing similar content on their YouTube stream. The discussion host had to gloss over the negative reactions in order for them not to be obvious to someone watching the broadcast of the Q&A session. However, if one was watching live on YouTube, it was possible to see the variety of comments and questions that appeared on the live comment board, which diverted attention away from the rigorously directed broadcast. There was a sharp contrast between the formalised broadcast and scholarly-style panel discussion and the stream of spon-taneous responses that bubbled up from the web chat.

One of the Performance Room curators, Catherine Wood, later recounted the sense of discomfort caused by this torrent of questions and comments, some of which were considered to be irrelevant to the performance and the overall purpose of the Q&A. She mentioned, "…it felt like putting art out in the Wild Wild West, in this no-man's land" (Pringle et al. 2014). Although the audience's reactions during the live streaming indicated that people were indeed watching the event live with a desire to interact, the disoriented comments often took attention away from the broadcast itself. The way that Wood refers to the extraterritoriality of the network is indicative of the challenge that the museum faced when staging live art online. The online space of YouTube is presented as an unknown territory; a limited portrayal of the platform as something that the museum cannot fully understand, which fails to recognise that YouTube is another type of institution itself – a "24/7 global archive" as the media theorist Richard Grusin has described it (2009, p. 66) – with its own "visitors" and systemic decisions behind its operations.

The main concern of the curators was their responsibility towards the artists and the performance piece, so there was a desire to limit the risk of negative attention associated with this exposure on YouTube. Although Tate had used YouTube before, mainly as a channel and archive for TateShots and recorded events, the use of the platform as a means of primary access to live art was totally new. And despite the fact that the performance was presented inside the framework of Tate's channel on YouTube, there was still the impression that the artist and his work were exposed to a community of people whose behaviour differed from the audiences that frequented the traditional museum spaces.

Tate's initial desire to stage a live performance programme online in a way that would experiment with new boundaries and open up the museum to new audiences soon transformed into a more defensive position under the perceived need to control this experience and contain the audience interaction. The first Performance Room event provided the impetus for a "protected space" where these types of online experiments could take place, as it was described by the programme curator (Pringle et al. 2014). The contrast between the stream of spontaneous audience responses and the formality of the live broadcast created concern amongst museum staff over the reception of the artwork and the artist's intentions. For that reason, the programming team suggested that the series should be further curated in order to avoid similar uncomplimentary incidents in the future (Zouli 2018, pp. 173–185).

Indeed, after the first performance, two main changes were applied in the programme format: the first was to simplify the broadcast and focus only on the live performance piece and the discussion with the artist. As a result, the host and the pre-recorded footage were removed so that the Q&A included only the artist(s) and the curator. These alterations indicated a turn towards a more immediate experience for the viewers, which corresponded both to the way the museum stages live art events in the physical space and to the documentation of interviews for the camera, as in the example of the TateShots series (see Figure 8.1).

The second change concerned the distributed audience and their role in the live event. After the unpredictable stream of responses of the first performance, the programme was then advertised and promoted only via mailing lists and specific cultural channels. The call for targeted viewers was a way to safeguard the artists' work by assembling an audience that is knowledgeable or has a specific interest in live art. Adding to that, the Tate producers started using a moderator control panel in the duration of the live online event. Through the panel, the moderators could choose which questions or comments under the hashtag #BMWTateliveQ were appropriate to appear on social media, including the YouTube comment

FIGURE 8.1 A sign in the corridor outside the Tate gallery room where the live broadcasting of Performance Room was taking place. Tate Modern, September 2014. Image courtesy of Capucine Perrot

board.[3] A more organised content management system was useful in order to avoid another situation in which unanticipated and inappropriate comments disrupt the visual experience and impact on the reception of the performance (see Figure 8.2).

In the case of Performance Room, Tate's desire to participate in the digital moment and welcome new audiences in online settings is tempered by the need to control the production of art and the formation of knowledge. The balancing between new technological practices, art historical traditions and strategic guidelines is complex and difficult to succeed. From the moment that museums became "mediatic,"[4] an important contradiction arose between the desire to use media technologies as a way to democratise the museum experience for the audience and the need to apply practices of control in order to secure the reception of this experience (Henning 2006, p. 74). The case of Performance Room prompts us to consider the museum's decisive power not only on staging an art piece through online broadcasting technologies but also on who is appropriate to host and participate in a discussion around art in this context.

In his analysis of Performance Room, the performance scholar Philip Auslander (2016) describes the liveness of watching the performances online in terms of two interconnected dimensions. Firstly, the "broadcast liveness" that bases the

FIGURE 8.2 The backstage of the live performance broadcast. One can notice on the front table people looking at the screens that broadcast the live footage from the camera(s) in the room; on the table on the back, the laptops of the Tate Media moderation team. Tate Modern, May 2014. Image courtesy of Capucine Perrot

experience of a mediated, live, event exclusively on the temporal co-presence of the viewers and the performer(s) setting aside the spatial co-presence, which is usually necessary for the experience of the "live" (Auslander 2016, p. 111). The broadcast liveness describes the migration of television to online landscapes, which has manifested mainly through the development of streaming services and the distribution of on-demand content; what Ramon Lobato has defined as "television's on-going metamorphosis into an online medium" (Lobato 2016, p. 13). The first part of the Performance Room broadcast could be described as Tate's live television on YouTube; however, the second part of the stream is what takes this description further. Television's convergence in the networked screen image is a condition that, according to Andrew Dewdney, "calls forth new conditions of reception, new subject positions, which need to be understood" (2013, p. 109).

The second dimension of liveness proposed by Auslander relates to an alternative mode of spectatorship, compared to the established mediations of performance art, which he calls "Internet liveness." The Internet liveness of Performance Room is based on the agency of the audience to engage with this live event as an active choice amongst other online activities. As the media theorist Geert Lovink discusses, multitasking is the basic quality of the contemporary media experience and total attention is unachievable; when one watches YouTube, other windows are naturally open and other things happen outside of the screen (Lovink 2008, pp. 11–12). The act of watching and commenting on the live Performance Room stream is, therefore, considered part of a "self-curated flow" of navigation and attention positioning (Auslander 2016, p. 123).

The element of Internet liveness is both what Tate wanted to address through this project and what turned out to be most unprepared for. The majority of commissioned artists dealt with their Performance Room pieces as performances *for the camera*, creating impressive installations, dance pieces, lighting or costumes, yet paying little attention to the performative qualities of the networked medium involved.[5] YouTube was seen merely as a streaming platform, and while the Q&A sessions often addressed questions about the online nature of the programme, the dissemination of performance in the performative context of online interfaces was neither taken into consideration when designing nor when documenting Performance Room.

The liveness of the screen interface supplemented the Internet liveness of Performance Room. The space where the performances took place is not the gallery room in Tate Modern but the screen instead, Auslander highlights (2016, p. 115). For that reason, it was important to pay attention to what was happening in the YouTube interface while the performance developed. One of the core elements was the comment board that appeared adjacent to the video frame, which included a live flow of questions or comments by the audience. Despite the use of a control panel to moderate the questions that appeared both on YouTube and Twitter, the comment streams confirmed the presence of the audience and added a layer of interpretation to the piece that was spontaneous in its departure.

It was only on the final season of Performance Room, in 2015, that Tate activated the live chat feature on the YouTube page, which allowed viewers to chat with each other as well as with the page host during the live event without using hashtags. It is worth mentioning here that for this final year of the programme, the lead Tate Media producer for the online broadcasts was a freelance collaborator and not a staff member from Tate Media. This change gave a fresh approach to the programme and allowed for the easier adoption of new features. The producer was the voice behind Tate's YouTube profile during the broadcasts and she interacted directly – under Tate's handle – with the other participants in the chat (Zouli 2018, pp. 248–249). However, the live chat conversations were temporary and live-only features, so as soon as the live streaming ended, they disappeared. There is also no trace of them today since they were not documented or recorded in any way.[6]

As a consequence, the Performance Room documentation as it can be found archived on Tate's YouTube channel, includes only the video broadcast of the performance and the Q&A session; hence, there is no evidence of the live question flow that was integral in the development of the online experience of the performance throughout the duration of the programme. Without the inherent element of the audience's texture on the live event, the archived videos became a combination of performance art documentation and an interview in a talk show setting.

In the afterlife of the programme, the video frame is being given priority over the screen interface. The museum dealt with the Performance Room video recordings as performative artworks combined with a live discussion event, both contained within the shared space of the video frame. It seems that in order to manage the unexpected risks and challenges that emerged from the encounter with online audiences, the museum transformed the programme into something that it could contain and control, namely a piece of video content on its channel or an object in its collection.

In 2016, as part of the opening of the new Tate Modern, it was possible for visitors to watch the Performance Room commissions on screens at the foyer of Tate Tanks. In this context, the only way that the audience could recognise that the videos were part of a live streaming event online was only by listening at the Q&A discussion at the end of each performance and not from any other elements in video or the screen.[7] The displacing of the performances from the online space inside the museum created a new context from which to watch these works and, in so doing, transformed them into *documents* of live art. This relocation of the work into a physical space and its recontextualisation as a video installation effectively remodelled these performances into art objects within Tate's art historical tradition. In this sense, Tate was finally able to contain Performance Room in its display culture and create a protected space in which the performances could be viewed. In the years that followed, Performance Room was absorbed into Tate's digital archive, while the umbrella programme of BMW Tate Live continues until today with events performed, and enjoyed by audiences, in the physical spaces of the museum.

Tate could have considered an alternative documentation of the programme that would retain all the distinguishable features of liveness in the process. Such an expansive documentation of performance works would also require, though, according to Dekker, Giannachi and van Saaze (2017), a re-consideration of the concept of the museological document as something more complex that could incorporate "physical and digital attributes, as well as visual and textual documentation" (p. 62). Ultimately, it would mean that the museum recognises the performative elements of the networked interface and the participation of the audience as integral parts of the performance piece. As Alexander Galloway describes it, an "interface is not a thing but an effect (…) an interface is always a process or a translation" (2008, p. 939). It would be therefore interesting to see museums embracing the processual qualities of the network and be more open to computational structures as operative agents. On the whole though, in the case of Performance Room, the anxiety of what happens when a performance becomes an image embedded in software on a screen prevailed over the initial curiosity about distributed cultural programming.

The conditions imposed by the coronavirus pandemic could have been an opportunity for Tate to revisit what was learned from the Performance Room experience. After the closure of the museum in March 2020, all the live performance events at Tate Modern were cancelled. As an immediate response, Tate invited its audience to watch "an online-only performance" by the artist Faustin Linyekula whose work was originally among the commissions for that year's programme of the BMW Tate Live Exhibition (Tate 2020). Since access to physical spaces was not allowed, the museum did a one-off staging of the artist's work, which was performed for the camera in the empty space of Tate Tanks. The promotion of the screening as an *online-only* performance highlighted that online access was the only way to watch the performance. At a moment when Internet liveness – to return to Auslander's idea – became habitual for audiences in lockdown, Tate remained in the schema of one-way transmission and broadcasting culture. Preserving control over the production of cultural knowledge in online settings remains a priority for the museum. This recent example from Tate's use of the Internet for art programming further provokes questions of whether the network is still seen as the 'wild west' by art museums; as an untamed territory that is a space of potentiality as well as risk.

When the Online Is the (Only) Alternative

After the first shock of the Covid-19 outbreak and the application of lockdown measures, it became clear that online platforms would form the main channels of communication, entertainment and social life until shelter in place orders subsided. During the first lockdown, live streaming and the circulation of video amplified. Film festivals relocated their screenings online; theatres and performance venues broadcasted live performances or archived performances of older works; and museums broadcasted new and old exhibition-related content such as interviews, live talks and tours. These practices positioned the museum in the role of the cultural entertainer, producer and/or educator, which is the way museums' online channels have been operating for more than ten years now.

For curators Marco De Mutiis and Jon Uriarte (2021), the pause of the museum activities in the physical space and the potential migration to online spaces was a chance for institutions "to rethink how collective networked spaces could be utilised for the exchange of culture and art, as well as an opportunity to consider the function and role they perform". The total shift of attention towards the online, which the conditions of the pandemic forced, was indeed a unique moment in the history of public art institutions that had to continue working outside the statement architecture of their gallery spaces. However, the reality was far from the reflexive moment that De Mutiis and Uriarte had envisioned. The lockdown left many institutions at sea and the difficulty to disconnect their work from the physical facilities, collections and pre-decided programme became evident. The museum experience transferred to a personalised setting in the viewers' computer screens either through virtual tours or by staging online events such as artists' and curators' talks, roundtables and learning programmes via YouTube, Zoom or other video conferencing platforms. As a result, the concept of "bringing the museum to you" through the hashtag #museumfromhome became popular among art institutions, as De Mutiis and Uriarte highlight.

The simulation of the museum in the digital sphere did not move beyond its ordinary framework of operation and instead mirrored pre-established analogue practices. Digital became, according to Ivanova and Watson's analysis, "synonymous with accessibility and reaching greater audiences to experience events designed for physical spaces (e.g. via streaming) or gallery-confined artefacts (e.g. via digitising collections)" (Ivanova and Watson 2021, p. 37). The use of video content as means of cultural entertainment not only mirrored older practices of broadcasting but also pointed to the fact that cultural organisations had to compete for the audience's attention with online streaming platforms such as Netflix or Twitch, with gaming platforms, or online communities such as Reddit (Ivanova and Watson 2021, p. 14). The pace and extent to which museums adapted their practices and art programming to the "networked normal" of the pandemic were disproportionate to the capacity of other industries to remodel their practices.

Twitch, for example, saw a 20% increase in its traffic in the first months of the pandemic (Koeze and Popper 2020), while by the end of the first year of the pandemic, it had an 80% growth in users' hours watched (Stephen 2021). Twitch appeared in 2011 as a live gaming streaming platform, and by 2017, it had more than 2 million unique broadcasters per month and about 10 million daily active users. The platform introduced what the sociologist T.L. Taylor (2018) has described as a "new form of networked broadcast": a live streaming where the broadcasters could interact with their audience in real time through a live chat and the interactions were directly integrated into the live event. Apart from game streams, the platform has been used widely for sharing, live or on-demand, creative work, concerts and music events, arts and crafts shows, cooking or "social eating" (Taylor 2018, p. 6). Twitch's networked broadcast resembles what Tate aspired to do with Performance Room – a combination of television, online spaces and networked communication. At the time when Performance Room programme launched, Twitch already existed as a platform; however, YouTube was more established as a channel for museums' video content. During the pandemic, online gaming saw an immense increase with large communities of people watching video content and online databases for billions of hours every month.[8] However, very few cultural organisations engaged with Twitch[9] or other online platforms that audience communities already occupied.

The Serpentine curators Victoria Ivanova and Kay Watson (2021) suggest that the next step for cultural institutions should be to engage with advanced virtual worlds and invest in the users' experience of art in hybrid spaces. This provocative suggestion, influenced by the technological and social potential of game industries and targeted to contemporary networked audiences, premises that art museums – or at least some of them – will overcome their "fear of the digital"; a fear that, according to Maria Chatzichristodoulou (2013, p. 313), is both metaphysical (fear of the "alien" characteristics of technology) and practical (fear of the transformational character of technologies which develop in high speed compared to the slower progress of the museum sector). In addition, an important step before testing virtual worlds is for museums to address online audiences as users. As it was underlined in the Performance Room example, and as the period of the pandemic indicated, art museums pay little attention to the users and their experience of interfaces when it comes to online exhibitions or the online broadcasting of art. Acknowledging the user is not only a way to address the attention economy in which online platforms are entailed but also a way to manifest the architecture and computational processes that host museums' online practices. As the net artist Olia Lialina (2012) notes, "being a user is the last reminder that there is, whether visible or not, a computer, a programmed system you use".

This is a step that could take time, as the pandemic showed that the relation of art museums to networked technologies remains complex and underemphasised in the level of both research and practice. Museums find it difficult to see the network as relational in reciprocal ways and have yet to explore networked culture as part of historical, political and technological processes.

Notes

1 According to a survey of 600 museums by the Network of European Museum Organisations (NEMO), more than 75% of the respondent organisations mentioned that during the pandemic they had "either increased their existing social media activities or started new" (Network of European Museum Organisations 2021, 4). Despite the remarkable drop-in visits and loss of income, the museum sector developed digital services as a response to the unexpected crisis. Social media platforms were already a space that museums used for the promotion of their activities, so Facebook and Instagram became two of the main ways of engaging with online audiences during this period. Adding to that, 53% of the respondent museums mentioned that they started or expanded the creation of video content.

2 Tate consists of four art galleries: two in London (Tate Britain and Tate Modern), one in Liverpool (Tate Liverpool) and one in Cornwall (Tate St. Ives).

3 The live audience could pose their questions using the hashtag #BMWTateliveQ on Twitter or Google+ as well as add comments on the Tate Facebook page under the relevant programme posts. The Tate social media team collected these responses in real time through the content management backend system and decided whether they were suitable for further posting on the YouTube comment board and/or queue them for the live Q&A between the curator and the artist(s).

4 The concept of the "mediatic museum" has been described by the media theorist Michelle Henning as the moment when museums started employing media and engaging with information technologies in their practices (Henning 2006).

5 Two examples of performances that embraced and responded to the networked context of Performance Room were that of Nicoline van Harskamp in 2013 and Selma and Sofiane Ouissi in 2014. Van Harskamp's work *English Forecast* broadcasted live on Tate's YouTube channel on 19 September 2013. The piece was "part performance, part participatory exercise" (Gormley 2016): the artist invited a group of four actors to recite a sequence of English words pronounced in different international accents, while during the live act, the audience at home was invited to repeat and record the word or sound that was being said/shown on the screen in their own accents. They were then encouraged to post these recordings on social media and, if interested, to even send them to the artist via email for her to incorporate them in her future work. The performance *Les Yeux d'Argos* by the Ouissi brothers took place on 18 September 2014 and explored "connection and disconnection, presence and absence in the digital realm" (Epps 2016). Sofiane performed

from the McAulay Gallery at Tate Modern, while a live video of Selma transmitted into the space from her apartment in Paris over Skype. The pair spoke and danced together through the monitor in the gallery and the online audience could see them both on the YouTube screen.

6 One can find indicative screenshots of these live-only features in my PhD thesis (Zouli 2018, 260–261).

7 The screen installations at the Tate Tanks included a headphone set for viewers to listen to the video of the broadcast. It is, however, doubtful that the average visitor would have spent approximately 30 minutes watching the whole of the video.

8 I refer here to Geert Lovink's (2008, 9) popular phrase that "We no longer watch films or TV; we watch databases".

9 Some examples of projects that used Twitch are *Screenwalks* by Fotomuseum Winthertur and The Photographers' Gallery or MoMA's exhibition of Ian Cheng's work *Emissary Forks at Perfection* (2021).

Bibliography

Auslander, Philip. 2016. "The Liveness of Watching Online: Performance Room." In *Perform, Experience, Re-Live*, edited by Cecilia Wee. London: Tate Publishing, pp 111–126.

Bishop, Claire. 2012. "Digital Divide: Contemporary Art and New Media" *Artforum*, September. www.artforum.com/print/201207/digital-divide-contemporary-art-and-new-media-31944

Chatzichristodoulou, Maria. 2013. "New Media Art, Participation, Social Engagement and Public Funding." *Visual Culture in Britain* 14 (3): 301–318. https://doi.org/10.1080/14714787.2013.827486

Dekker, Annet, Gabriella Giannachi, and Vivian van Saaze. 2017. "Expanding Documentation, and Making the Most of 'the Cracks in the Wall'." In *Documenting Performance: The Context & Processes of Digital Curation and Archiving*, edited by Toni Sant. London: Bloomsbury Publishing, pp 61–78.

De Mutiis, Marco and Jon Uriarte. 2021. "Curating the Pandemic Image." *LUR*, 26 May. https://e-lur.net/investigacion/curating-the-pandemic-image

Dewdney, Andrew. 2013. "Curating the Photographic Image in Networked Culture." In *The Photographic Image in Digital Culture*, edited by Martin Lister. London: Taylor & Francis Group, pp 95–112.

Dewdney, Andrew. 2019. "The Networked Image: The Flight of Cultural Authority and the Multiple Times and Spaces of the Art Museum." In *The Routledge International Handbook of New Digital Practices in Galleries, Libraries, Archives, Museums and Heritage Sites*, edited by Hannah Lewi, Wally Smith, Dirk vom Lehn, Steven Cooke. New York: Routledge, pp 68–91.

Dewdney, Andrew, David Dibosa, and Victoria Walsh. 2011. *Tate Encounters: Britishness and Visual Culture*. London: Tate Publishing.

Dewdney, Andrew, David Dibosa, and Victoria Walsh. 2013. *Post-Critical Museology: Theory and Practice in the Art Museum*. London: Routledge.

Epps, Philomena. 2016. "Selma and Sofiane Ouissi, Les Yeux d'Argos 2014." *Performance at Tate: Into the Space of Art*, April 2016. www.tate.org.uk/research/publications/performance-at-tate/case-studies/selma-and-sofiane-ouissi

Fei, JiaJia. 2013. "Digital as a Dimension of Everything." *Medium*, December 19, 2013. https://medium.com/@vajiajia/digital-as-a-dimension-of-everything-120c86c4dab5

Galloway, Alexander R. 2008. "The Unworkable Interface." *New Literary History*, 39 (4), Autumn: 931–955. www.jstor.org/stable/20533123

Gormley, Clare. 2016. "Nicoline van Harskamp, English Forecast 2013." *Performance at Tate: Into the Space of Art*, April 2016. www.tate.org.uk/research/publications/performance-at-tate/case-studies/nicoline-van-harskamp

Grusin, Richard. 2009. "YouTube at the End of New Media." In *The YouTube Reader*, edited by Pelle Snickars and Patrick Vonderau. Stockholm: National Library of Sweden, pp 60–67.

Hein, Hilde S. 2000. *The Museum in Transition: A Philosophical Perspective*. Washington, DC: Smithsonian Books.

Henning, Michelle. 2006. *Museums, Media and Cultural Theory*. Maidenhead: Open University Press.

Higgins, Peter. 2015. "Total Media." In *Museum Media*, edited by Michelle Henning, 305–326. Chichester: Wiley Blackwell.

Ivanova, Victoria and Kay Watson. 2021. *Future Art Ecosystems: Art x Metaverse*. London: Serpentine Galleries. https://futureartecosystems.org/

Koeze, Ella and Nathaniel Popper. 2020. "The Virus Changed the Way We Internet." *New York Times*, April 7, 2020. www.nytimes.com/interactive/2020/04/07/technology/coronavirus-internet-use.html

Lialina, Olia. 2012. "Turing Complete User." *Contemporary Home Computing*. http://contemporary-home-computing.org/turing-complete-user/

Lobato, Ramon. 2016. "Introduction: The New Video Geography." In *Geoblocking and Global Video Culture*, edited by Ramon Lobato and James Meese. Amsterdam: Institute of Network Cultures, pp 10–24.

Lovink, Geert. 2008. "The Art of Watching Databases: Introduction to the Video Vortex Reader." In *Video Vortex Reader: Responces to YouTube*, edited by Geert Lovink and Sabine Niederer. Amsterdam: Institute of Network Cultures, pp 9–12.

Maculan, Lena. 2008. "Researching Podcasting in Museums: Can New Broadcasting Models of Publication Make Art More Accessible?" PhD diss., University of Leicester. http://hdl.handle.net/2381/4015.

Network of European Museum Organisations. 2020. *Survey on the Impact of the COVID-19 Situation on Museums in Europe: Final Report*. Berlin: NEMO. www.ne-mo.org/fileadmin/Dateien/public/NEMO_documents/NEMO_COVID19_Report_12.05.2020.pdf

Network of European Museum Organisations. 2021. *Follow-up Survey on the Impact of the COVID-19 Pandemic on Museums in Europe – Final Report*. Berlin: NEMO. www.ne-mo.org/fileadmin/Dateien/public/NEMO_documents/NEMO_COVID19_FollowUpReport_11.1.2021.pdf

Pepi, Mike. 2014. "Is a Museum a Database?: Institutional Conditions in Net Utopia". *e-flux Journal,* 60, December. www.e-flux.com/journal/60/61026/is-a-museum-a-database-institutional-conditions-in-net-utopia/

Pringle, Emily, Catherine Wood, Rebecca Sinker, Derek McAuley, Mark Miller, and Leyla Tahir. 2014. "Online Collectives: Do Online Collectives Constitute a New Public Space for Museums?" *Panel Discussion*. London: Cultural Value and the Digital: Practice, Policy and Theory. June 2, 2014. www.tate.org.uk/context-comment/video/cultural-value-digital-audio-series

Richard, Birgit. 2008. "Media Masters and Grassroot Art 2.0 on YouTube." In *Video Vortex Reader: Responces to YouTube*, edited by Geert Lovink and Sabine Niederer. Amsterdam: Institute of Network Cultures, pp 141–152.

Ringham, Jesse. 2011. "Tate Social Media Communication Strategy." *Tate Papers*, 15. www.tate.org.uk/research/publications/tate-papers/15/tate-social-media-communication-strategy-2011-12

Snickars, Pelle and Patrick Vonderau. 2009. "Introduction." In *The YouTube Reader*, edited by Pelle Snickars and Patrick Vonderau. Stockholm: National Library of Sweden, pp 9–21.

Stack, John. 2010. "Tate Online Strategy 2010-2012." *Tate Papers*, 13. www.tate.org.uk/research/publications/tate-papers/tate-online-strategy-2010-12

Stack, John. 2013. "Tate Digital Strategy 2013–15: Digital as a Dimension of Everything." *Tate Papers*, 19. www.tate.org.uk/research/publications/tate-papers/tate-digital-strategy-2013-15-digital-dimension-everything

Stallabrass, Julian. 2013. "The Branding of the Museum." *Art History*, 37 (1): 148–165. https://doi.org/10.1111/1467-8365.12060

Stephen, Bijan. 2021. "Twitch and Facebook Gaming Exploded during the Pandemic—And They're Even Bigger a Year Later." *The Verge*, March 15, 2021. www.theverge.com/2021/3/15/22331623/twitch-facebook-gaming-pandemic-hours-watched

Tate. 2011. "Tate and BMW Announce Major New International Partnership: BMW Tate Live." Press. www.tate.org.uk/about/press-office/press-releases/tate-and-bmw-announce-major-new-international-partnership-bmw-tate

Tate. 2012. "Jérôme Bel – BMW Tate Live: Performance Room." YouTube Video, 47:23. March 23, 2012. www.youtube.com/watch?v=l0TmUQmKpDg

Tate. 2013. *Tate Report: 2012/2013*. London: Tate Publishing. www.tate.org.uk/download/file/fid/55015

Tate. 2019. "BMW Tate Live." Exhibitions & Events. www.tate.org.uk/whats-on/tate-modern/performance/bmw-tate-live

Tate. 2020. "BMW Tate Live Exhibition: Our Bodies, Our Archives." Exhibitions & Events. www.tate.org.uk/whats-on/tate-modern/exhibition/bmw-tate-live-exhibition-2020

Taylor, T.L. 2018. *Watch Me Play: Twitch and the Rise of Game Live Streaming*. Princeton, NJ: Princeton University Press.

The Met. 2011. "Connections." YouTube Playlist. April 9, 2021. www.youtube.com/playlist?list=PL42CF2A007311B93C

Vayanou, Maria, Akrivi Katifori, Angeliki Chrysanthi, and Angeliki Antoniou. 2020. "Cultural Heritage and Social Experiences in the Times of COVID 19." In *Proceedings of AV I 2CH 2020: Workshop on Advanced Visual Interfaces and Interactions in Cultural Heritage (AV I 2CH 2020)*, Ischia, Italy, September 29, 2020. http://ceur-ws.org/Vol-2687/paper2.pdf

Vlassis, Antonios. 2020. "Online Platforms and Culture: The Winning Actors of the Great Lockdown?" *Global Watch on Culture and Digital Trade*, 4 May. https://ficdc.org/wp-content/uploads/2020/05/MAY-2020-_-N%C2%BA4.pdf

Walsh, Victoria. 2016. "Redistributing Knowledge and Practice in the Art Museum." *Stedelijk Studies*, 4. www.stedelijkstudies.com/journal/redistributing-knowledge-practice-art-museum/

Walsh, Victoria and Andrew Dewdney. 2017. "Temporal Conflicts and the Purification of Hybrids in the 21st-Century Art Museum: Tate, a Case in Point." *Stedelijk Studies*, 5 (Fall). https://stedelijkstudies.com/journal/temporal-conflicts-and-the-purification-of-hybrids/

Walsh, Victoria, Andrew Dewdney, and Emily Pringle. 2014. "Modelling Cultural Value within New Media Cultures and Networked Participation." Final Research Report, The Royal College of Art, London South Bank University and Tate. www.rca.ac.uk/documents/432/Cultural_Value_and_the_Digital_Programme.pdf

Zouli, Ioanna. 2018. "Digital Tate: The Use of Video and the Construction of Audiences." PhD diss., London South Bank University. https://doi.org/10.18744/PUB.002751

9

SCREENSHOT SITUATIONS

Imaginary Realities of Networked Images

Magdalena Tyżlik-Carver

Introduction

To answer the question "what do pictures want?," W. J. T. Mitchell, in his book of the same title, organises his response into three parts, each dealing with what is considered to constitute a picture: image, object and medium. Mitchell defines the image as referring to "any likeness, figure, motif, or form"; the object is "the material support in or on which an image appears, or the material thing that an image refers to" and is objecthood understood as something "set over against a subject" (2005, p. xv). The medium therefore produces the picture through the means of material practices bringing image and object together. The logic of this structure asserts pictures as "complex assemblages of virtual, material and symbolic elements" (Mitchell 2005, p. xiv).

This chapter considers the proposition that pictures take part in worldmaking in Nelson Goodman's sense of images that "produce new arrangements and perceptions of the world" (cited in Mitchell 2005, p. 93). If a picture "refers to the entire situation in which an image has made its appearance" (2005, p. xiv) the screenshot might be helpful as a unit of measure of such a situation in the networked context. Screenshot as a unit of measure is a material phenomenon that while capturing what is on the screen gives evidence to how it is assembled, and how its virtual, material and symbolic elements *matter* in the world. Screenshots, when taken as units of measure, participate in processes of knowing, measuring, theorising and observing, and as such they are "material practices of intra-acting within and as part of the world" that help in learning "about specific material configurations of the world's becoming" (Barad 2007, pp. 90–1). Rather than following representationalist[1] belief that representations are ontologically distinct from what they represent, screenshots give account of how these phenomena are dependent, intra-actively

DOI: 10.4324/9781003095019-13

rearranging and (re)producing not only images but also situations necessary to generate screenshots.

In popular culture, screenshots are images of the screen that capture what is on it at a particular moment in time. This documentary character makes them especially suited to register histories of technologies (Allen 2016; Gaboury 2021; Gerling 2018), art (Rossenova, Chapter 11 this volume), and broader social and cultural practices that take place across networks (Gaboury 2019; Soon & Tyżlik-Carver forthcoming). While many discussions of screenshots focus on links with photography, their networked character is less often addressed. This volume takes up the question of what the networked image might be and this chapter takes a screenshot as format that reveals the networked character of such an image.

Today, screenshots and their situations happen in millions while interactions with computers have long been part of daily lives. Screenshots are reproduced in unprecedented numbers while they are also established as technical and cultural practices. The contemporary computational apparatus that facilitates generating screenshots is distributed and networked and available worldwide to billions of users who own smartphones and computers and who can immediately capture and share such screen images. While screenshots continue to circulate as images, they require infrastructural support. Distributed with them are metadata and other material practices that involve technologies, skills, interfaces and habits, often in form of gestures that are performed on and with the help of computational devices. These are screenshot situations which, in the tradition of non-representational practices, ask us to look beyond what is visible. Screenshots are units that define screenshots as situations which reproduce, distribute and circulate more than images.

The proposition of this chapter is that images are world-making phenomena with agential force and screenshots are its material instances. This force, however, cannot be measured as an image's effect on viewers, which could be represented on a scale, let's say from 1 to 5. Rather, the agential force of an image can be defined as a distance – in time and space – between an image and its apparatus, for this distance demonstrates the extent of the image's range across networks: the operations that are executed and required to propel it across this infrastructure. The moment of making an image, which in this case is the event of screen capture, initiates a process of executions that require specific socio-technological conditions that reproduce screenshot situations computationally, and the duration of which becomes measurable as distance. On one hand, what is on the screen is contained within a picture frame, and on the other, even if this particular image has been captured already, what remains is the continuous possibility of the situation of capture. Barad's emphasis on measurement as "a meeting of the natural and the social" and as "a potent moment in the construction of scientific knowledge" (2007, p. 67) applies here: the distance and number of reproductions across material networks, the executions of which are social and computational constitutes a measurement of socio-material potency.

Two approaches in particular guide my making sense of screenshots as world-making phenomena, one which belongs to the history of technology and another driven by artistic practice. From the perspective of design and architecture histories, screenshots have been thought of as instrumental to the development of computer-aided design in the 1960s (Allen 2016). Matthew Allen discusses the application of screenshots for architecture by situating them "in contexts of practice" and as "figures of knowledge" that define the practice (Allen 2020, p. 103). Rosa Menkman, in contrast, researches the materialities of computer screens and specifically resolution architectures and technologies that render data into images. These two authors usefully locate images in relation to practices, devices and formats created—that is, to the social, historical and technical conditions of image-making.

These approaches both apply to two artworks from the Screenshots: Desire and Automated Image, an exhibition that I curated in 2019 in the Gelleri IMAGE in Aarhus, Denmark. The works "Drawing with Sound" (2017) by Anna Ridler and "The Conceptual Crisis of Private Property as a Crisis in Practice" (2003) by Robert Luxemburg are chosen specifically because they inform the idea of a screenshot as a condensation of many different material and computational practices across time and space. These works, both of which use the "screenshot," are relational images that capture the computational processes and give evidence of the apparatus of their making. This self-disclosure of the image's production and distance of travel reconfigure the screenshot's potency as a measure of what a network image is.

Screenshot Situations

Let us start by looking at the environments where screenshots appear. To print, grab or shoot a screen engages algorithmic operations, storage capacities, image files and formats, as well as screens, and graphical and hardware interfaces. Indeed, such environments are interactive and need to be operated, and today humans, bots and other algorithms might act as such operative agents. To address environments where images appear, Mitchell proposes that they are given a face, or a figure. The advantage is said to make media (imaginatively) accessible as "landscapes or spaces" (2005, p. 208) to picture not just what media are and what they do but also "to raise the question of what the medium of theory itself might be" (p. 209). While such an approach might embody media theory by embedding it in media practices, it simply replaces one set of figures with another. This is especially the issue in digital spaces, which many users don't see as landscapes that might open horizons, but impenetrable black boxes, the internal operations of which are a mystery. The challenge is to account for what is happening on this terrain and what already have taken place in such a way that, while accounting for the lives of images, metaphors are not the only ways to access what we know about an image.

When looking at the aesthetic features of screenshots in order to typify images created with early graphical software, examples of which are discussed in the next section of this chapter, it is possible to locate screenshots' many lineages, with roots in science, drawing and photography as well as computational processing and programming languages, rather than seeking to approach their appearance metaphorically. These different histories and practices feed into what in the end is captured on the screen and how. Today it is less common to photograph computer screens in such a way that the whole context of image-making is in the picture, so it is possible to see, at least to some degree, how the image is made. Screenshot infrastructures are no longer visible because they have become part of the networked and computational infrastructure of which the screenshot is also a part. What drops out of the picture is the very possibility of situating screenshots as relational and connected to technological, social and historical conditions of their making. Seeing screenshots as landscapes and locating them as spaces can mitigate this to a degree by opening screenshots' metaphorical horizons. However, framing screenshot as a landscape, charge it with a particularly passive character that instead of locating an image as an intra-active phenomenon locks it into what it represents, immobilising also its interactive possibilities.

Based on material structures and social practices that generate screenshots, they are expansive environments of informational architectures and social, professional, aesthetic and technical practices, which are affectively entangled into a screenshot situation. As such screenshots never settle, even if represented in an image, as they are always liable to reproduction elsewhere. Screenshots are a vernacular mode of screen culture and they indicate which specific material processes mattered and how, at the moment of taking a screenshot. They actively rely on screen resolutions, compression algorithms and other computational processes specific to image-making infrastructures at a specific moment. One might ask what happens to the image, be it drawing or photography, if the tools of its making are replaced with a computational device? Or, to approach this question differently: how do the aesthetics of screenshots change when human-computer interaction is an everyday phenomenon in the communicative practices of social media use?

Histories of Screenshot Situations

As Winfried Gerling (2018) shows, screenshots' history can be traced back to screen image photographs of cathode ray tubes in the mid-twentieth century and even to earlier photographs of X-ray screens by Brazilian doctor Manuel de Abreu. In the mid-1930s, Abreu developed a method to capture X-ray film with a small-format camera, making it possible to screen thousands of patients for tuberculosis for a fraction of a price (Gerling 2018, pp. 150–1). This example illustrates screenshots' representational and functional roles to capture images of cathode ray tubes on oscilloscope or X-ray representations of patients' internal organs. The application was possible with techniques and technologies "born out of the massive growth in the electrical sciences during the nineteenth century" (Gaboury 2021, p. 64). While early screenshots originate in professional settings, they also belong to technologies and techniques, the genealogies of which can be linked to "wider range of media built around the display and manipulation of electrical currents" pointing to scientific rather than cinematic significance of early images of screens (2021, p. 63).

While the proliferation of screenshots might be synonymous with the contemporary ubiquity of screens in general, as well as wide familiarity with the photographic format made possible with digital camera (Rubinstein and Sluis 2008), I am interested in tracing moments when the scientific and vernacular come together, thus accounting for the heterogeneity of screenshots as always scientific and technical, while also social and aesthetic. Especially relevant to this line of thought is Mathew Allen's study of screenshots as "central to the task of constructing a new meaning for the computer" (2016, p. 638). Traditionally, drawing was a regularly used method to visualise concepts and ideas in design and architecture. However, in the early CAD programs, "screenshots conjured an unusual situation" (Allen 2020, p. 111), demonstrating the potential of interactive computational process and in this way helped to promote the idea that a computer can be more than just a calculating device (Allen 2016, pp. 638–9). More than an image format, screenshots demonstrated the novel use of a computer with a graphical interface and they represented human-computer interaction influenced by J. C. R. Licklider's vision of human-computer symbiosis as a "cooperative interaction between men and electronic computers" ([1960]; Licklider 2003, p. 75). With screenshots, it was possible to *show* that such interactivity between humans and computers is possible and what it might look like (Allen 2016).

Sketchpad, a software developed at MIT by Ivan Sutherland as part of his doctoral research in 1963, was highly influential in developing the notion of interactivity that is now familiar to any user of computational devices. Sketchpad was one of the first graphical user interface (GUI) programs designed to create an environment for manipulating the graphical elements displayed on the screen of the computer with a light-pen, which allowed the user to modify them by pointing, clicking and dragging. This kind of interaction is a commonplace today, and most usually facilitated by the use of the computer mouse or fingers on the touchscreen. At the time, computers were seen as massive calculators used at specific moments to perform specific tasks while being operated by people working in computer labs. Sketchpad was developed for TX2, an experimental digital computer at MIT used for designing and testing and it could only work with this one computer (Sutherland 2012). The program used drawing as "a novel means of communicating with the computer" (Sutherland 1963), making it one of the first graphic systems that used language only for legends, and the first non-procedural programming system.[2] Demonstration of how the software works to the audience was necessary also in order to show what kind of interaction is possible, and that such interaction is already a reality (Allen 2016, p. 641). Rather than communicating with the computer through written commands via computer terminal, Sketchpad's model of interactivity used drawings to graphically visualise the processing necessary to execute the program.

Let us remember that the first screenshots were indeed photographs of computer screens taken by a camera which was not part of the computer. There were practical reasons for this. On one hand, computer monitors were even less common than computers, and on the other, computers simply did not have the capacity to process or store such images, not to mention that algorithms for formatting images did not yet exist. A number of different kinds of screenshots are included in Parts of Sutherland's doctoral thesis, show Sutherland drawing with the light-pen on the computer display, while others display drawings made with the computer, or attempts to capture and visualise computational process behind the image, such as a CRT (Cathode Ray Tube) diagram of architectural drafting (Allen 2020). Between 1960 and 1963, according to Allen, at least four different conventions converged into a screenshot understood as "an image type that represents the interactive computer and the process of using one" (2020, p. 111). Such screenshots not only show "particular things (this or that drawing of a geometrical object), but, more importantly, they give a second-hand impression of the *experience* of using an interactive computer" (2016, p. 658 my emphasis). In effect, as well as representing computer-generated drawings, screenshots documented the whole situation of making such an image, which included a human operator, the machine, as well as tools facilitating input and output.

These early screenshots, or screenshot *situations*, modelled forms of interactions that are standard today by visualising a certain concept of interactivity between humans and computers. In the context of Sutherland's presentation of the Sketchpad, this situation involved a number of elements such as a model of a computer for which the software was specially designed; the program was operated by the software designer, who was the only person able to use the software; Sutherland's work was supported by research programme in an internationally recognised institute with research facilities unlike any other in the world. All these elements constituted a very special and expensive research infrastructure that was not widely available. To get the first screenshot, an infrastructural support of machines, software, people and research funding was necessary to generate the situation of taking a picture of a screen. Screenshot situations are not just about what they represent—the images captured on the screen – but about the very situation of their creation at the moment of interacting with the computer.

Situated Screenshots and Their Resolutions

What are the infrastructures that make visualisations of computational processes possible? The screen, monitor or other forms of digital display frame a screenshot as an image usually captured with the PrtSc, or some other combination of keys, depending on the kind of the device (PC or mobile phone) and model. From memory boards to image compression algorithms such as JPG or PNG formats, to the actual hardware of screens, they all define what and how it becomes part of the image. However, because a screen is generally considered synonymous with an image it is rarely theorised independently (Gaboury 2021, p. 55).

Such theorisations do exist, even if as exceptions to this tendency. One example is the project *institutions of Resolution Disputes* (i.R.D), by artist and theorist Rosa Menkman. Working from the vantage point of artistic research practice, Menkman engages the materiality of computer screens and histories of digital image formatting to "uncover how resolutions inform both machine vision and human perception" (Menkman 2020). According to the artist, "resolutions do not only impose how or what gets run or seen, but also what images, settings, ways of rendering and points of view are forgotten, obfuscated, or simply dismissed or unsupported". Menkman traces the construction of digital images along five axes: habit, materiality, genealogy, institutions and scale. This orientation of imaging technologies and techniques places an image in relation to these presumably distinct phenomena and shows how they are continuously entangled. What becomes apparent in this methodological approach is that screen is understood as multidimensional environment that combines the technical specificity of a computer screen with deeply social, gendered and racial conditions of contemporary culture and aesthetics that have been part of making the digital image. The project shows how screen is deeply relational and technical, and resolution is taken as a unit with which to measure relational and not just technical affordance of the screen.

Menkman's research develops a particular form of resolution forensics, a method that reconstructs by creating "the recreation-backwards of the event from the trash and the rubble that it leaves behind" (Weizman 2015, p. 37). Motivated by the desire to "refuse syntaxes of history to direct our futures," i.R.D methodically inquires into experiences of digital spaces as resolved with dimensions, surfaces, lines, frequencies, vectors and connections between them as they render and are rendered into a digital image, and as they organise visual representations and their objects. This project reveals how these technologies omit and leave behind some things while underlying and making visible others. The problematic of a digital image is mapped onto the issue of the presumed objectivity of such technical images, which Menkman questions using her five-axiom method.

Colour test cards, originally developed by Kodak in the 1950s, exemplify such presumed objectivity. Menkman uses them as objects for her resolution forensics to reveal the standards for colour calibration in photographic image. These Shirley cards or Shirleys, named after Shirley Page, the first Kodak studio model for the prints, and other versions such as China Girls or Lena are "the artifacts of obsolete film history" (Menkman 2020), which have been absorbed as part of a standard deeply rooted in bias towards lighter skin tones, a major issue also present in contemporary machine vision technologies.[3] The assumption of the objectivity of technology is a major concern when it comes to the application of these formats as they set standards by seeping into the everyday through discreet and highly distributed use. Habit, materiality, genealogy, institutions and scale are helpful axis on which to locate colour test cards to trace their technological, scientific, social and cultural dimensions.

Menkman's research as part of *institution of Resolution Disputes* identifies the problem of presumed objectivity of resolution technologies. Images like Shirley cards are example of a meta-image, an image about images, as they themselves are protocols and standards of colour calibration. If objectivity is the aspiration "to knowledge that bears no trace of the knower" (Daston and Galison 2007, p. 17), the research of i.R.D project shows how this takes place through automation that absorbs decisions into technological features, and how certain ideals that are gendered, racialised and objectifying and not objective are coded and calibrated into colour standards. And while these standards are distributed and highly visible, the process of coding slowly disappears until it is completely transparent and difficult to trace. If screenshot is an image and a social and technological process, screenshots give an entry into socio-technological situations of making an image that is networked and locatable in specific infrastructures where presumed objectivity can be traced to certain subjects.

Desire and Automated Image

Allen's histories of screenshots and Menkman's resolution disputes show how world views are represented and how they circulate as images. The visions of human-computer interaction and colour calibration standards are located in these examples giving evidence of computational technologies and how they develop at certain points in time and specific circumstances, and bodies that these technologies capture differently. Screenshots and screens are used to account for these histories. The exhibition Screenshots: Desire and Automated Image (2019) also takes computational image represented in a screenshot to see what is part of such an image. The curatorial question asks what happens to the image when a camera is replaced by a computer and computational process? Ten works were included, each involving a version of a screenshot as an aesthetic, computational and often networked format.[4] Dictionary definitions of images describe them as physical likeness, representation or optical counterpart, but they also refer to mental representation of something previously perceived, in the absence of the original stimulus, referring to an idea and conception. Already this definition points to the curious relation between the presence of the object and also its absence, and image is in constant state of crisis captured by and in the image: the thing represented in the image is not the thing itself (see Cox, Chapter 5).

This gap between the thing and its representation is precisely what can be measured with the screenshot. And as this chapter argues, what you get with the screenshot is more than you can see.[5] Construction of images is increasingly automated and screenshots give a measure of this process by giving account of the relation between the image and its apparatus by looking at the screenshot situation in its entirety, that is, by linking social, technological, historical and aesthetic conditions of image-making. The exhibition Screenshots: Desire and Automated Image investigates the screenshots potential as a measure of what is in the image because its focus is on how artists engage and experiment with highly networked and automated technologies for image-making, and how they use affordances and constrains of these technologies to offer solutions and speculate on possibilities of image-making technologies as ways to imagine with. The two image-works analysed below are specific examples of how screenshot as a unit of measure is helpful in identifying situations that define affordances and constraints of created images.

"The Conceptual Crisis of Private Property as a Crisis in Practice"

I start with the work *The Conceptual Crisis of Private Property as a Crisis in Practice*[6] (2003) by German artist, film-maker and programmer Robert Luxemburg (aka Sebastian Lütger) to account for the changing conditions not just in image-making but specifically in understanding what image becomes. The image is of a screenshot of a Mac computer desktop taken at 11:57 am, on Tuesday, 20 April.[7] Within the main frame, there are familiar icons, logos, menus, images and five more application windows open. But while Luxemburg's work is a picture in a frame that can be hung on a gallery wall, it simultaneously exists as a series of digital files: crisis.php (a source code), crisis.txt (a text file explaining the work) and crisis.png (an image), all part of the complete artwork (Figure 9.1).

The artwork is an example of software art, an art format where software is not just a tool to make art with but where it becomes an artwork itself. This is perfectly illustrated with this image which is also software that when inserted into a computer executes a program generating the ASCII version of a novel *Cryptonomicon* by Neal Stephenson. *The Conceptual Crisis of Private Property as a Crisis in Practice*

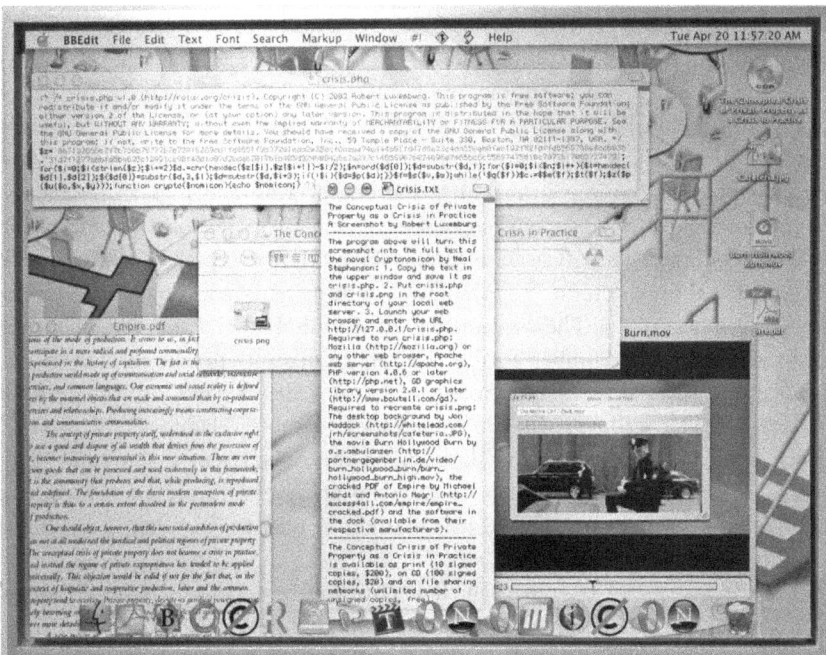

FIGURE 9.1 Robert Luxemburg (2003) *The Conceptual Crisis of Private Property as a Crisis in Practice* (exhibition view 2019, photo by Geert Skaerlund. Courtesy of Galleri Image and the Artist)

practically hides an encrypted version of the novel and in this way it exists as both an original artwork by Luxemburg and, if executed, an illegal copy of a literary work by another author. As a result a provocation that indicates conceptual crises resulting from the fact that we are dealing here with a digital format of an image, its affordances and properties that change the very assumptions about objects of art and culture, and practices which generate and govern them. This provocation is directed at everyone, the computer user, gallery visitor, copyright lawyers, art world and computational systems, and it asks what kind of object is a digital image, who does it belong to, and what does it represent? What can be done to the digital image and who/what can do it? Using conceptual steganography, based on the cryptographic practice of concealing messages in another message or an object, Luxemburg plays with its possibilities as conceptual artistic practice and asks: "What would an image that contains a novel actually show? What would the code look like? Could the code be the image, and the image the code? And so on...."[8]

The artwork makes the viewer wonder: how to read this image? The title already points towards the crisis of property, but we can see that the image is in some other crisis too. A first look at it shows that it is over saturated with elements that are more or less visible in the picture. Nothing is clearly represented: each element demands that we look closer at it to see what exactly is in the picture. The string of recognisable icons on the bottom of the screen reads "cryptonomicon" referencing the title and subject of Stephenson's novel, while suggesting that Luxemburg's image too is cryptic and that things are hidden in it. The least visible is the background image, a wallpaper using one of Americam artist Jon Haddock's isometric drawings. Isometric projection is a way to visually represent three-dimensional objects in two dimensions in technical and engineering drawings, and Haddock uses this technique to create representations of cultural and historical events in the style of a computer game. A series of twenty 16-bit drawings prepared by Haddock in 2000 for the exhibition No Absolutes[9] depict scenes and objects which are both real and fictional. The assassination of Martin Luther King and Theodore Kaczynski's (the "unabomber") cabin are interchangeably presented next to fictional events such as the picnic from the movie *Sound of Music* or a scene from *Godfather II*. The wallpaper in Luxemburg's artwork uses Haddock's "Cafeteria," which depicts a scene from the shooting in the cafeteria in Columbine High School in 1999. There are other visual elements present in the picture, which have to be seen as separate first before it is possible to understand what is assembled together to create this image.

The foreground of this image includes five different open browser windows, each an example of a different digital format and a digital object in its own right. One of them is a PDF (Portable Document Format) version of the book Empire by Antonio Negri and Michael Hardt, open on the fragment that discusses the changing condition of private property. The title of the artwork is taken from this fragment. There is also .mov file Burn Hollywood Burn by a.s. ambulanze, next to three other windows displaying the code, an explanation how to run it, and a PNG version of this very screenshot. In all, this desktop image is a curious mise en abyme, which in this case tells stories of images by hiding them in other images. In doing so, not only is it more difficult to understand its meaning, but the very concept of what the image is blown up by amplifying its digital properties through unlimited distribution of the image via file-sharing networks as well as 10 signed physical prints and 100 signed CDs.

The Conceptual Crisis of Private Property as a Crisis in Practice is a noise of formats and representations that all seem to be equally important and brought to the front to capture our attention. As they are all assembled together into an image and as an image, they also create its depth. Formally, assemblage is a three-dimensional construction, a form of a collage often created with found objects that are assembled into a "thing," an installation, or a sculpture. At first sight *The Conceptual Crisis of Private Property as a Crisis in Practice* hardly seems to be a collage of any kind as it is a digital print of a desktop screenshot. It is not much different from what is traditionally expected from an image, a representation superimposed onto a surface and nested within a frame. It is at this point that the viewer is challenged to look beyond the visual and focus on computational functions inserted in the image. The screenshot as a unit of measure maps the image as a multidimensional space between the virtual materiality of a computer, symbolism of visual culture, and wide-spanning effects of digitalisation. What is assembled exists across dimensions which are differently available to algorithms and to people.

This work in itself represents a desire to understand what a digital image is by rendering its features also computationally. The fact that this image can be executed and that it contains other executable formats moves it beyond the representational dimension of what normally is defined as an image. This software image reconstructs computationally the image's potentials and virtual possibilities that are just a moment away from being executed. And at the same time, it is a representation, a mirror on what an image is computationally.

Drawing with Sound

The question of what images are and what they represent, remains. Anna Ridler's work *Drawing with Sound* (2017) helps to unpack critical dimensions of images further. Developed in collaboration with sound artist and programmer Ben Heim, this performative work challenges the very idea of image, and, we might add, the screen or screenshot as capturing computational processes. The work exists at least in three formats. First it is a performance of a drawing with charcoal on a white painted panel, which is executed by the artist who wears a specialised headset, a set of glasses equipped with a web-camera connected to a computer (Figure 9.2). The computer runs a neural network algorithm which translates lines of the drawing picked up by the headset camera into data for the computer to process. These data are added to a growing dataset and they are used as an input data source which the algorithm outputs as sound. These sounds are immediately performed during the drawing and they are also recorded. Once the performance is finished the work exists in the

FIGURE 9.2 Anna Ridler (2017) *Drawing with Sound* (performed during the opening 2019, photo by Mikkel Kaldal. Courtesy of Galleri Image and the Artist)

exhibition as a documentation of a drawing on the panel, and a sound recording— more of a snapshot of a process, than a screenshot representing what is on computer screen. As in the case of Sutherland's Sketchpad, Ridler is presenting to us another model of interacting with the computer and its algorithms; we are challenged to look at the whole situation. The work demonstrates human–computer interaction, which together with the screen, the drawing, the camera and the sound, are part of the apparatus that makes the image. A differently networked image.

Drawing with Sound gives us a snapshot of a computational instrument that connects and networks together the wooden panel on a wall, artist drawing on it, the drawing body equipped with camera-glasses, neural network algorithm, dataset and sound compositions. They are all part of this computational instrument. As the artist moves filling the picture frame with lines and other figures, with her eyes equipped with headset glasses, she registers these figurations and passes data onto a computer. Here the drawing is a verb and a noun, an act of drawing and drawing as an object to be drawn and looked at by the artist and her machine. At the same time, this is an act of creating a dataset, and simultaneously playing the instrument as each line becomes a source of data that can be output as sound.

Drawing as a performative act creates an image and, at the same time, generates and expands a dataset for training neural networks used for this and future versions of the work. Neural networks, an algorithmic model for data processing, use a t-distributed stochastic neighbour embedding (t-SNE) method, a statistical method for visualising high-dimensional data by locating each data point on two-dimensional or three-dimensional map. The original dataset used for training the model was based on the collection of artist's drawings made over two years. The drawings were scanned and then broken into grids to work out the most common marks to be matched with sonic data. And as drawings are translated into data distributed and networked in the abstract computational space, neural networks perform calculations able to predict and plot data which can be output into an infinite number of sound compositions.

Source data is necessary to start the work and neural networks generate ever more combinations and models, which while not visible to human eye create ever new representations. Inke Arns notices how "the technical structures that observe and act performatively have increasingly withdrawn into invisibility." and how these performative structures "are recognisable only by their effects but are no longer necessarily visible" (Arns 2010, p. 261). While Arns writes about performativity of interfaces, this description is accurate for neural network models too, which are invisible to humans and outside of their full control. Neural network calculations as predictions generate layers of computational relations that remain in the computer until their output in form of sound utterances.

Like in Luxemburg's work, here image is also activated. Ridler's drawing becomes part of the computational process, with its operative function becoming "part of an operation" (Farocki 2004, p. 17). Such a function is characteristic of our "post-optical age" of complete transparency understood not as "simplicity, clarity and controllability through viewability" but its opposite characterised by "invisibility and information concealment" (Arns 2010, p. 256). This is what Menkman reveals in her research into screen resolution. But rather than remaining invisible and enclosed for computational operation (Arns 2010, p. 262), *Drawing with Sound* captures action as a cybernetically modelled interaction between a machine and human body. They conjure the feedback loop as an ensemble of the sound and drawing as they are performed and executed together. If Sutherland's Sketchpad demonstration functioned as a source of images to represent and model human and machine interactions, *Drawing with Sound* models circulation of data between neural network and the artist.

Ridler's work might be read as a poetic interpretation of how human–machine interaction can be imagined anew in the era of data extractions and dataveillance. Screenshot, according to Allen, always stands for something more polished, and in the case of *Drawing with Sound*, as in the case of Sketchpad demonstration, the generated drawing is part of modelling of other ways of human and machine communication. In this case, it focuses on data collecting that is affirmative, creative, and based on mutual (human and machine) exchange of information and not its extraction. *Drawing with Sound* is an environment and testing ground with screenshot incorporated as part of human and computer interaction for post-optical age.

But to consider the image in the *Drawing with Sound* as a trace of human and machine communication would be only part of the story. We also see how it is possible for the image to change and morph into multidimensional forms that live across physical and latent spaces as they are drawn by the artist and immediately represented within the virtual 3D space that can be read by a computer and then output into a sound format. Screen panel is an operational unit that when studied closely can give insight into image as human and computational phenomenon, not focused on an object but giving space to interaction and collaboration. At the same time, while Ridler's gaze is captured and directed at the drawing (the verb and the noun), it performs shifting boundaries of such an image, and how what is not visible is incorporated back into a picture even if output as sound value. Here the tension between presence and absence, that is, the stuff of desire and also stuff of visual order, governs images' aesthetics. Transparency of automation, in this case invisibility of data processing that happens within neural networks, is experienced as output of sound interrupted by gestures of artist's body moving her hand while affected by aural experience. Image rather than limited by the picture frame continuously rearranges itself into relational situation between the artist and machine, which together take part in worldmaking and not representing and mirroring it.

Conclusion – Image Logistics

Screenshots are not surface images capturing what can be seen on the screen. Rather they contain image logistics, infrastructurally managed and historically conditioned flow of relations between traditions, practices, technologies, bodies and formats. It is in these logistics that networked character of images lies. Screenshots discussed here are confusing as they construct a picture as imaginary "pulsating with life" and "suggesting the motion and vitality that is captured, even in the still image, the material excess that turns the virtual image into a real picture" (Mitchell 2005, p. 74). In both cases, we can see images, either a drawing or a screenshot of a desktop, but each of them is only a part of a fuller picture. It is with a screenshot as a unit of measure that it is possible to account for what is in the picture beyond what we can see. The challenge is to read an image while conceptualising its apparatus and screenshot as unit of measure helps in responding to such a challenge. In case of Luxemburg's work, such an apparatus executes the image as code and novel, and in *Drawing with Sound*, it is performative and based on human and machine collaborations.

Conversely, screenshots are situations that capture the circulation of data at the moment when data materialises as an image. The real conditions of screen capture become an image, and they are prototyping techniques used to capture computational possibilities and not just to represent what can be seen on the screen. Screenshots help to measure the distance between computational possibility of the apparatus and representations visible on the screen as part of apparatus operations. My interest in screenshots is practical, namely to use it as a unit of measure, which can reveal the networked character of such images as a set of material effects. The statement that all images are networked might seem trivial; however, it is in the deconstruction and unpacking of these images that it is possible to specify what kind of material and computational networks are contained within the image.

Using the example of screenshots, I position the networked image not as a new category of images but rather as a specific and located agential force – a situation that organises relations and bodies that come together to make an image.

Notes

1 I use here Karen Barad's definition of representationalism as "the belief in the ontological distinction between representations and that which they purport to represent; in particular, that which is represented is held to be independent of all practices of representing" (2007, 46).
2 This is how Alan Kay defines Sketchpad in the video recording of him commenting over the original Sketchpad demo by Sutherland. See this video recording (Carroll 2007).
3 See the work of Algorithmic Justice League and projects such as Gender Shades (Buolamwini and Gebru 2018).
4 I engage here with two of these works. For details of the other artworks in the exhibition, see the exhibition plan https://bit.ly/3pEXaBQ. Images from the exhibition are available at https://bit.ly/354DJt1].
5 See also Lozana Rossenova (Chapter 11, this volume), who defines screenshots also in regard to data and metadata describing screenshot images, thus making them accessible online. Rossenova locates screenshots networked potential in "relations between backend, frontend and user agent."
6 This work was nominated at transmediale festival in 2004 in category of Software Art.
7 No year is included with the date.
8 Full interview with the artist available at https://rolux.org/texts/interview/.
9 See Arizona State University Art Museum page for more details on Haddock's project and the exhibition https://asuartmuseum.asu.edu/content/screenshots-project-jon-hadd ock-exhibition-no-absolutes. See also this blog post on the Screenshot project by Jon Haddock https://argotandochre.wordpress.com/2009/09/01/isometric-screenshots-the-art-of-jon-haddock/.

Bibliography

Allen, Matthew. 2016. "Representing Computer-Aided Design: Screenshots and the Interactive Computer circa 1960." *Perspectives on Science* 24 (6): 637–68. https://doi.org/10.1162/POSC_a_00227.2020.

Allen, Matthew. 2020. "Architecture Becomes Programming: Invisible Technicians, Printouts, and Situated Theories in the 1960s." In *The Figure of Knowledge: Conditioning Architectural Theory, 1960s–1990s*, edited by Sebastiaan Loosen, Rajesh Heynickx, and Hilde Heynen. Leuven: Leuven University Press, pp 101–26.

Arns, Inke. 2010. "Transparent World Minoritarian Tactics in the Age of Transparency." In *Interface Criticism: Aesthetics beyond the Buttons*, edited by Christian Ulrik Andersen and Soren Pold. Aarhus: Aarhus Universitetsforlag, pp 253–76.

Barad, Karen. 2007. *Meeting the Universe Halfway: Quantum Physics and the Entanglement of Matter and Meaning.* Durham: Duke University Press.

Buolamwini, Joy and Timnit Gebru. 2018. "Gender Shades: Intersectional Accuracy Disparities in Commercial Gender Classification." In *Proceedings of the 1st Conference on Fairness, Accountability and Transparency*, 77–91. PMLR. https://proceedings.mlr.press/v81/buolamwini18a.html.

Carroll, David. 2007. *Sketchpad, by Dr. Ivan Sutherland with Comments by Alan Kay*. www.youtube.com/watch?v=495nCzxM9PI.

Daston, Lorraine and Peter Galison. 2007. *Objectivity*. New York and Cambridge, MA: Zone Books; Distributed by the MIT Press.

Farocki, Harun. 2004. "Phantom Images." Translated by Brian Pool. *Public*, no. 29, 12–22.

Gaboury, Jacob. 2019. "Screenshot or It Didn't Happen." *Still Searching: Fotomuseum Winterthu.* www.fotomuseum.ch/de/2019/07/15/screenshot-or-it-didnt-happen/.

Gaboury, Jacob. 2021. *Image Objects: An Archaeology of Computer Graphics*. Cambridge, MA: MIT Press.

Gerling, Winfried. 2018. "Photography in the Digital." *Photographies* 11 (2–3): 149–67. https://doi.org/10.1080/17540763.2018.1445013.

Licklider, Joseph Carl Robnett. 2003. "Man-Computer Symbiosis." In *The New Media Reader*, edited by Noah Wardrip-Fruin and Nick Montfort. Cambridge, MA: MIT Press, pp 74–82.

Menkman, Rosa. 2020. *Beyond Resolution*. i.R.D. https://beyondresolution.nyc3.digitaloceanspaces.com/Rosa%20Menkman_Beyond%20Resolution_2020.pdf.

Mitchell, William J. T. 2005. *What Do Pictures Want? The Lives and Loves of Images*. Chicago, IL: University of Chicago Press.

Ridler, Anna. 2017. *Drawing with Sound*. Performance with machine learning algorithm.

Robert Luxembourg. 2003. *The Conceptual Crisis of Private Property as a Crisis in Practice*. Inkjet print.

Rubinstein, Daniel and Katrina Sluis. 2008. "A Life More Photographic." *Photographies* 1 (1): 9–28. https://doi.org/10.1080/17540760701785842.

Sutherland, Ivan E. 1963. "Sketchpad: A Man-Machine Graphical Communication System." PhD, Lincoln Laboratory: Massachusetts Institute of Technology. https://web.archive.org/web/20130408133119/http://stinet.dtic.mil/cgi-bin/GetTRDoc?AD=AD404549&Location=U2&doc=GetTRDoc.pdf.

Sutherland, Ivan E. 2012. "The TX-2 Computer and Sketchpad." *Lincoln Laboratory Journal* 19 (1): 82–4.

Weizman, Eyal. 2015. "Forensic Temporality." In *Simulation, Exercise, Operations*, edited by Robin Mackay. Falmouth: Urbanomic, pp 37–40.

PART IV

Digitisation and the Reconfiguration of the Archive

10

NETWORKS OF CARE

Annet Dekker

The Paradox of Digital Sustainability

Since the advent of digital technology, cultural heritage has been produced, stored and preserved in digital form by cultural producers and heritage institutions. This has resulted in a large body of what is called "digital heritage." Digital heritage – whether singular, born-digital art projects or large-scale digital humanities projects – is constructed of different technical layers and is characterised by multiple human and machine processes. Consequently, its continuation, and thus preservation as the activity by which it is kept functional, relies heavily on technical equipment and sociopolitical infrastructures. The complexities and challenges of preserving digital heritage can be summarised as follows: reading older code and software can be difficult; obsolete technology and the reliance on third, often commercial, parties pose problems; software and hardware maintenance can be very time-consuming and expensive; with different people working on a project, changes to projects appear over time; and art projects in particular can evolve into other versions, which makes it hard to define what an art project is or consists of in the digital environment (Dekker 2010; Depocas 2003; Hodge 2000; Rinehart and Ippolito 2014). While the digital preservation practice described in this chapter is specific to cultural heritage and art, the complications, challenges and solutions are also relevant to understanding the wider issues of networked culture, which is characterised by a similar assemblage of human and non-human actors.[1]

In the past two decades, several solutions to preserve digital heritage have emerged (Dekker and Falcão 2017; Engel and Wharton 2014; Rechert et al. 2013). While some of them work well, in many cases the content and information changes, as most hardware and software follow the economic model of planned obsolescence (Fitzpatrick 2011; Pope 2017). Consequently, endless migration, emulation, virtualisation and documentation tools and projects are being set up to prolong

DOI: 10.4324/9781003095019-15

the functioning of digital heritage. However, a focus on high-end technical preservation methods for maintaining digital heritage is revealed to be unsustainable and questionable. This happens at the level of the method: preservation approaches such as migration, emulation or virtualisation risk changing the form and content of projects, and similarly, with every software upgrade the media environment in which these projects exist can further change their aesthetics and functioning (Dekker 2018; Rinehart and Ippolito 2014). Consequently, specialist knowledge and expertise are also continually required to solve new technical challenges and at the same time, non-professionals who are engaged in preservation efforts will need specific guidance, both of which are a burden to most organisations (Summers 2020). Finally, the enduring technical rat-race comes at a high energy cost, which results in significant carbon footprints for digital heritage projects, and thus digital preservation presents a challenge to the ecological environment (Bhowmik 2019; Cubitt 2016; Gabrys 2011). Taken together, a tension emerges between the need to keep digital heritage safe for future research, cultural memory or evidence, and the continuing need to update technical tools and methods to enable these art projects to survive but which poses an increasing burden on organisational infrastructures and methods as well as on the ecological environment. In other words, digital sustainability is a preservation dilemma, or even a paradox.

In recent years the literature about digital sustainability has resonated in digital heritage organisations, where sustainability is mobilised to improve gallery spaces, and waste and energy management to minimise the ecological footprint (De Silva and Henderson 2011; Kagan 2011; Pendergrass et al. 2019; Tansey 2015). As a result, many organisations set their environmental goals by directing their attention to financial and staffing resources. At the same time, overwhelmed by the constant technical changes, several artists have decided to delete their projects. For instance, in 2011 Slovenian net art pioneer Igor Štromajer ritually deleted a number of his classic art projects that were produced between 1996 and 2007; because of changes to technical settings and the updating of the web, the art projects no longer looked or functioned as he had once intended (Sakrowski 2017). While Štromajer prefers deletion to aesthetic loss and malfunctioning; in other cases, users have started to take care of decaying art projects (Rinehart and Ippolito 2014; Van Saaze 2012; Zavala et al. 2017). In such instances, networks emerge wherein tasks and responsibilities are distributed and shared. I termed such networks "networks of care" (Dekker 2015, 2018). Here the challenge of preservation shifts from the object itself to the maintenance of a network that supports the art project (Laurenson and Van Saaze 2014).

Museum and conservation studies have a long-standing and valuable perspective on preservation but have been slow to respond to the potentiality of involving expertise from beyond their realm (Wharton 2011). In general, they have avoided the topics of social process and cultural change and how these could affect the sustainability of digital heritage, mostly as these may challenge institutional values and processes (Nowvisky 2019; Prelinger 2019; Rinehart and Ippolito 2014). Moreover, by focusing on the uniqueness of the object and its technical aspects they neglected the importance of the complex and inherently changing sociotechnical infrastructure in which digital art projects thrive. Sustainability, in the sense of preservation, is as much a problem of governance as it is of technical and environmental constraints. In this chapter, I move beyond the economic or quantifiable benchmarking of sustainability and emphasise another potential area in which digital preservation can become more sustainable: by focusing on the potential of networks of care as a way to preserve an art project. While these processes happen in all types of arts, they are particularly manifest in net art projects, because in those cases there are strong relational dependencies between different technologies, people (artists and users), and ideas that cooperate in the realisation of an art project.[2] It will become clear that such an approach is not merely a material or technical solution to fix a project, as many net art projects – similar to a networked and relational image-assemblage – are embedded in and develop as part of a sociopolitical and technical environment that will need to be taken into account when considering their preservation.

Care as a Conceptual Device and Practical Method

In 2011, digital humanities and media studies scholar Kathleen Fitzpatrick suggested that the preservation of digital objects may become less about "new tools than new socially-organized systems, systems that take advantage of the number of individuals and institutions facing the same challenges and seeking the same goals" (Fitzpatrick 2011). Similarly, in 2014 Head of Collection Research Tate Pip Laurenson and researcher Vivian van Saaze concluded in relation to preserving performance art in the museum that: "It is not the problem of non-materiality that currently represents the greatest challenge for museums in collecting performance but of maintaining – conceived of as a process of active engagement – the networks which support the work" (Laurenson and Van Saaze 2014). Even though the importance of thinking about preservation within a network structure that consists of social relations is gaining traction, it is important to note that the network is not yet seen as inherently part of an art project. Instead, the network is seen as facilitating a project or a preservation approach. Yet, what happens if the network is considered as an actor rather than a tool? Moreover, in what way could such a network be said to care?

Evidently, the notion of care is very present in preservation practice: collection care is pretty much at the heart of preservation. Yet, here I want to try to move beyond caring for an object, and instead focus on care as a relational practice. While the concept of care is used and interpreted in different ways depending on academic or professional discipline, country and culture, I follow the notion of "care" as conceptualised and described by Annemarie Mol in her ethnography of health care.[3] In her book *The Logic of Care* (2008), Mol describes how care is not merely a matter of making well-argued individual choices but is something that grows out of collaborative and continuing attempts to attune knowledge and technologies. Care is understood as involving professionals and patients but also other material elements and technologies. Similar to how humans' responses can be ambiguous, the research shows how the unintended effects of technology can impact the course of care (Mol et al. 2010). In other words, they stress how care is relational: a set of heterogeneous practices that is local and specific and involves a "persistent tinkering in a world full of complex ambivalence and shifting tensions" (Mol et al. 2010, p. 14). Moreover, in care, the action is more significant than the actors: the latter may shift and change, but the relational actions remain important. While Mol makes explicit what it is that motivates care: an intriguing combination of adaptability and perseverance; feminist scholar María Puig de la Bellacasa emphasises how care is also never neutral. It is ambivalent, simultaneously necessary and oppressive, it suggests affect but also asymmetrical power relations; moreover, it provides space to think about possible worlds. In this sense, it is open-ended and invites (or provides space for) speculation (Puig de la Bellacasa 2017). Clearly, and as mentioned by anthropologists Mol and Hardon, "engaging in *caring* does not serve an unequivocal, common good. To think that it does is yet another romantic dream (Puig de la Bellacasa 2017). Caring practices, like other practices, are rife with tensions" (Mol and Hardon 2020). By using the concept of care as a tool to analyse the *activity of caring* that happens in preservation, and more specifically in digital art projects, I understand care as specific, situated and complex, yet also as a relational and processual activity that develops over time rather than being performed in a single moment. As Mol and Hardon point out, such an "activity of *caring* is not taken on board by isolated individuals, but spread out over a wide range of people, tools and infrastructures. Such *caring* does not oppose technology, but includes it" (Mol and Hardon 2020). Moreover, "The technology involved does not offer control, but needs to be handled with care – while, in its turn, it is bound to only work as long as it is being cared for" (Mol and Hardon 2020). I'd like to expand on this by emphasising the agency of technology *within* and *through* the network of care.

Framing a Network of Care

The concept of "network" has a long history and can mean different things in different disciplines and discourses. Here I loosely follow the description of Deleuze and Guattari that characterises a network – they use the term rhizome – as a system of non-hierarchical connections without clearly defined borders: a "rhizome has no beginning or end; it is always in the middle, between things, interbeing, *intermezzo*" (Deleuze and Guattari 2004[1980], p. 27). This means that a network is a dynamic system in which it is not apparent when or where a network starts, or who starts it, nor that its development can be predicted.[4] Such temporality in care is not unusual: even institutional and conventional preservation practices tend to happen at unpredictable moments – either when something breaks while on display, or when something is taken out of the depot due to a loan request and is then examined.

One of the conclusions that came out of my earlier research on the conservation of net art was that these preservation efforts often are maintained and/

or prolonged by different individuals, who collaborate as a network of care. By addressing these networks *as* care, I aim to draw attention to the significance of practices and experiences that are rendered invisible or marginalised by conventional and dominant "successful" – and mostly Western – forms of institutionalised preservation practices. From a pragmatic point of view, a network of care is based on a transdisciplinary attitude and a combination of professionals and non-experts who manage or work on a shared project. More specifically, for a network of care to succeed outside of an institutional framework it ideally has to consist of several characteristics. These can be identified by looking at how a network gives agency to the different actors involved.

To summarise, ideally a network of care adheres to a transdisciplinary attitude, consisting of a non-hierarchical or informal structure with different levels of expertise. To enable the creation and administration of a project, the transmission of information is facilitated by a common mode of sharing in which everyone in the group has access to all the documents or archives. Ideally this is an open system or a dynamic set of tools that is used and also cared for, where users can add, edit and manage information, and track changes. Such a system can also be monitored by the network, potentially both by the users and the machine itself. An added bonus is that if someone leaves, the project can continue because the content and information is always accessible and embedded in a larger network. This allows users to take control of a shared project, thus obtaining meaning from their "investments." To be able to share information and benefit from experience and insights gained elsewhere (for example, in other networks dealing with similar issues), a network should be dynamic, so that individuals can move easily between roles and projects, which can also be merged or divided among smaller or more specialised groups.[5]

While investigating the social sustainability efforts of several net art projects, I noticed different types of networks of care. While most emerge from urgent issues, or are formed around an emotional connection, they often develop and are organised in different ways. Here I made a distinction between how a network of care can be: (1) (part of) the art project; (2) an artistic preservation approach; and, (3) a proposition as part of a pilot study. Analysing these different approaches will highlight the challenges and potentials of a network of care for digital preservation.

Network of Care as Art Project

In 1997, Martine Neddam launched the website *mouchette.org*. The project presents Mouchette, a nearly 13-year-old French girl who lives in Amsterdam and speaks English and French. Mouchette uses the website to tell her personal life story on several web pages, which started to populate the site throughout the years.[6] Mouchette never ages, but visitors gradually learn more about her troubled past, as the website, and the project as a whole, expands into more and more web pages and projects, albeit that it is not entirely clear what the "whole" project actually is.

Neddam refers to the character Mouchette as a metaphor that she uses to create meaning around issues she finds relevant:

> It's hard to say what constitutes *mouchette.org*. Over the years I have lost track of all the performances, projects and objects that I made. But for sure, *mouchette.org* is more than just a website. (…) When I started Mouchette I wanted to use the notion of a character as something that transcends media, I saw the character as something that can be used as a form, or a container, this allowed me to gather and structure information. I have always believed that a character, a person or an identity is a good metaphor. They can assume the identity of an institution without actually existing. In this sense, I see characters as containers that carry units of meaning.
>
> *Dekker 2011, p. 22*

In 2003, as part of the mouchette.org project, Neddam started a Mouchette network (*mouchette.net*), an open platform where anyone could be Mouchette. Members can use Mouchette's identity to send e-mails, upload their own image to the main site or create their own version of a Mouchette website. This network grew over the years, and several versions of Mouchette appeared.

As a counterpoint, the *ihatemouchette.org.net* was started to support those sharing stories about why they don't like Mouchette. Besides being a communication and presentation tool, or a "social space" and "a platform of exchange," as Neddam refers to them (Dekker 2011), the networks are also intended to confuse people by

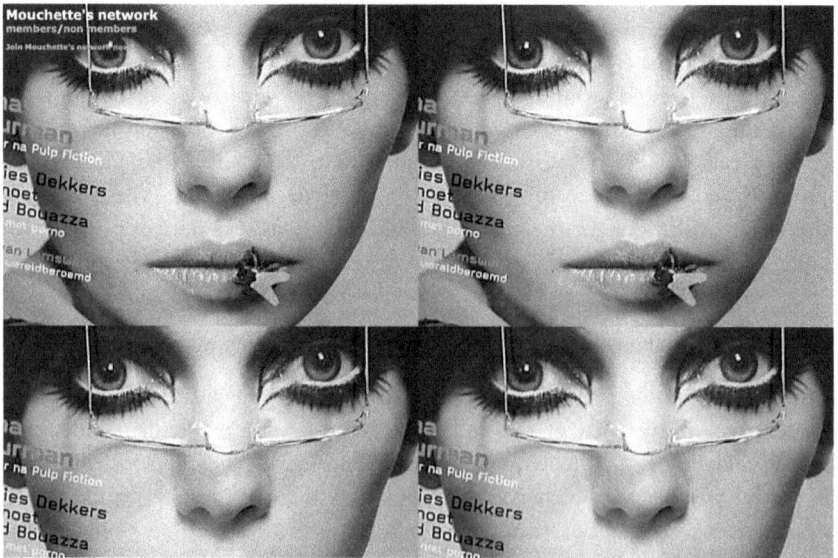

FIGURE 10.1 mouchette.net, screenshot

allowing users to create alternative or anonymous personas, and by not being clear about identity and authorship.[7] Similar to the anonymous character of *mouchette. org*, the identity of the *mouchette.net* members is not always clear, nor is their ownership of the project. Neddam herself mentions how she considers the additional sites as valuable versions of her Mouchette project (Black 2020). Although the idea of creating different networks is important to Neddam, she uses them primarily to reach out to and connect with the fans of Mouchette, and not (yet) as a preservation strategy.

Yet such a situation arose when *mouchette.org* was threatened with court action. In 1997 Neddam launched a quiz in which she compared different Mouchette characters: her own version against the main character in the film *Mouchette* (1967), directed by Robert Bresson – both characters are based on the novel *Nouvelle histoire de Mouchette* (1937) by Georges Bernanos. The widow of the French director was not amused by the comparison between the characters: mouchette.org was regarded by many as a controversial website, as it had become the topic of heated debates in the French news. Particularly her webpage in which she addresses the topic of suicide by asking what a suicide kit for children (as a toy to learn more about suicide and play "pretend suicide") should contain, and started a forum where people could respond and give advice, was not taken lightly. In 2002, a few years after the launch of the game, mouchette.org received a summons from Bresson's widow – reinforced by the French Society of Dramatic Authors and Composers (SCAD) – cautioning her to delete any reference to her husband's film from the website.[8] If not, more legal measures would be taken. Neddam decided to remove the French part of the game, but at the same posted the letter and the story on various e-mail lists. A chain-response followed in which several organisations and individuals – some of whom were part of the various Mouchette networks – said that they would mirror the game on their servers. Distributed via different servers and websites the French game can still be played.

A network of care emerged, both as an emotional response and to protest an urgent culturally (or politically even) unjust issue. As an informal structure, some of the organisations managed to preserve and still host the work while others changed direction and lost or deleted the project in the process. This is not uncommon and happens particularly with art projects that are not cared for by institutions (Van Saaze 2012). Yet the example highlights the necessity of connecting users who are, or become, partly responsible for the management and accessibility of content, in which different aspects of preservation can also be applied. In other words, a process of negotiation and re-questioning develops, which Renée van de Vall connected to a form of "middle grounding" (Van de Vall 2018), a process in which different viewpoints such as stories about details, private disclosures and generic or general statements alternate. Such a structure in which different perspectives come together ensures that a process of gradual acceptance takes place. In this sense, borrowing from Puig de la Bellacasa (2017), a network of care is not maintained by individuals in terms of giving and receiving, but by a cooperative disseminated force in which the complexity of the circulation of care is all-pervasive. At the same time, it shows how a network of care can dissolve, or change direction, perhaps to return at another time.

Rather than an obstacle, a network of care as (part of) an art project is inherently temporal and unexpected, yet tracing the art project's historical changes benefits from a trail, for example, version control or other documentation systems, that clarifies the decision-making processes.[9] Finally, with networks being integrated into the concept and structure of the project, a network of care stands a better chance of being activated when needed. Whether or not this will be successful remains to be seen, but a focus on relations of temporality and care contributes to an acknowledgement of alternative ways of thinking about preserving art projects that are processual and networked in nature, either in or beyond the institutional purview and towards a practice that is more inclusive of the networks that are at the core of the project.

Network of Care as an Artistic Preservation Approach

As mentioned in the introduction, in 2011 Slovenian net artist Igor Štromajer announced on Facebook that he was going to delete a large portion of his earlier projects: "If one can create art, one can also delete it. Memory is there to deceive." According to the artist, his projects didn't look the same anymore because settings had changed and the web had been updated (Sakrowski 2017). Burdened by never-ending technical changes, updates, migrations and the threat of obsolescence, Štromajer preferred deletion to aesthetic loss and technical malfunction. The project, aptly titled *Expunction*, raises questions about temporality, duration, access and availability on the web and how these processes impact cultural memory. While his action provoked concern and indignation on Facebook: "Igor!!!!!!! Can't you do something else to go through your mid-life crisis????!!!!!,"[10] Štromajer continued to delete his older projects. Yet, at the same time he documented the entire process: screenshots and texts of the projects, the reviews about them and the conversations around *Expunction* are now all saved on his website and hard drive. As part of the performance, and as a subsequent phase of the project, in 2016 Štromajer sent an e-mail to a selected group of his contacts. I was one of the recipients and the e-mail read: "Dear Annet, I'm sending you five files. Please put them somewhere safe. Thank you very much, Igor" (Figure 10.2).

I opened the files and saw two abstract cropped images, two gifs (one of someone sitting on a toilet and one of a roll of toilet paper), and an audio file of less than a second. Two years later I received another similar e-mail, this time a bit more descriptive, asking me to put the – now-encrypted – files in a safe place. Asking him about his practice, Igor explained that −*oμ4x* (minus mu four times) is a "performative action" in which over a period of several years, from 2016 to 2022, he asks a decreasing group of people to safeguard several files (a random selection from his earlier project *Expunction*).

Looking for other modes of distributing, sharing and experiencing the art that is trapped and compressed in the removed files, Igor is organising an emerging

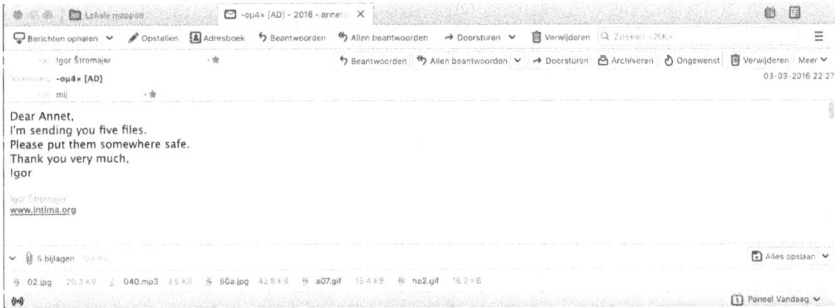

FIGURE 10.2 E-mail Igor Štromajer, *−oμ4x*, 2016, screenshot

network of guardians, or caretakers, oriented towards *becoming* rather than *being* (Harrison 2015, p. 27). The project proposes new modes of active engagement and creative use, and demonstrates an engaged way of dealing with circulation and relations, in which the distributive effects are intentional, even if what finally happens is unpredictable. Moreover, the repetition of the performative act of sending and receiving transforms the singularity of an affect into a sustained engagement. The extended period of waiting for something to happen affirms the reality of the events that unfold, even if the outcome is unknown. The project can disappoint but that is also its beauty: the potentiality of the event – the suspense or suggestion of infinity or of being part of an adventure, which may only become clear through engaging with it. Indeed, these images are hardly interesting by themselves, but together and as part of a larger whole they are compelling because they convey a suggestion of potentiality. As Štromajer suggests:

> It's a kind of a cycle, a durational, perhaps never-ending online performance with its natural rhythm: being constructed, deconstructed, then reconstructed anew, but this time differently. Who knows exactly what comes afterwards, but there is certainly no end to this cycle, because every trace, every move you make has its consequences.
>
> *cited in Sakrowski 2017*

There is no logic or predictability, and while the individual images and other files remain autonomous, all together and with the sparse e-mails they become networked images symbolising a promise, a proximity which one day may be fulfilled. In effect, the project feeds a continuous desire that keeps returning with each engagement. In addition to the technical files and encryption, which harbour their own technical specificity and agency, the social aspect of the act is important here.

−oμ4x reflects the complex temporality of many net art projects; arguably the network was instigated by the first e-mail, but its development is ambiguous, prone to rupture or loss, and the result is speculative, depending on the actions of the actors who don't know each other (yet). An example of what this could mean was

shown in 2019 when Bonnefanten Museum in Maastricht organised a solo exhibition of the Dutch process artist Ine Schröder. Schröder is known for allowing her art projects to "disappear," for giving them the opportunity to become something else, and many of her projects ended up being preserved by her circle of family, friends and acquaintances (Reinders 2019). Similar to the other examples I described, she did not regard her art as a sum of autonomous things, and she wasn't interested in something definitive or fixed; instead, she choose permanent transformation and described her process as a network of "staketsels," continuously reconfigured objects, connected in memory, space and time. Paradoxically, and not unlike Štromajer, she documented and archived her art projects meticulously. Discovering this archive after her death, the museum also stumbled upon a social network of "donors" who they contacted to find out what was left of the once given or discarded art projects. The documentary that was made of their discoveries shows how the individuals in the network care about the legacy. Each person assumed agency over the gift by paying close attention to the fragile constructions, gluing back pieces or rearranging the parts. Sometimes smothered with affection the projects were given new life. By presenting their art as public gifts, and choosing to circulate it among close friends instead of commercial or established artworlds, Schröder and Štromajer's approaches manifest networks that include points of convergence, yet these likely occur at an undecided moment when different actors find a point of connection, or shared interests, in which the roles of artist, audience, curator and conservator are allowed and sometimes encouraged to merge, leading to various and multiple narratives and solutions. Similar to the minute and multiple archival notes, instructions and documentation, a network becomes an invitation, a gesture to the future to continue a project that was never finished in the first place.

A Network of Care as a Proposition

When presenting some of the outcomes from my previous research on the concept of a network of care, I was often asked what it means to set up or become part of one, particularly from the point of view of an institution. Since I had focused on artists' projects, I also wondered what setting up and sustaining a network of care from an institutional perspective might involve? Art historian Karin de Wild and I initiated a pilot study to analyse the different actors within a network of care to learn more about their potential roles, and the benefits and challenges of setting up and sustaining a network of care. We selected the art project *Brandon* by Shu Lea Cheang. We had both worked with Cheang before and *Brandon* had been restored in 2017 by the conservation team at Guggenheim in collaboration with the computational department at New York University (NYU). The restoration was primarily focused on the recoding of the website and that other parts of and previous partners in the project were less involved in the preservation efforts. Our aim was to see whether it would be useful to form a network of care around the project to bring out the different aspects of *Brandon* by including these collaborators and developers. Such a network seemed relevant since Cheang in various interviews has emphasised that "*Brandon* is a multi-artist, multi-site, multi-institution collaboration" (Ho 2012), and from the beginning the idea was to keep the project growing.[11] We started by locating and talking to the main institutions who were actively involved in various stages of the art project's development to find out how they viewed their role in a (future) *Brandon*. These different stakeholders were important to comprehend the intricate nature of *Brandon*, which expanded and evolved beyond the main website, and to understand the relevance of *Brandon* at the time it was created, how it developed and the (historical) importance of the project today. So, what constitutes *Brandon*?

Cheang started *Brandon* in 1996 as a critique of social normality. *Brandon* was directly based on two articles that appeared in *The Village Voice* at the time: the court case around the rape and death of the 21-year-old Brandon from rural Nebraska in 1993, who was murdered for living life as a male despite being born female, and a notice about a rape spree that took place in a text-only chat room that left the victims feeling violated and bereft. The events touched upon some of the core themes in Cheang's work, in particular, the exploration of gender identities and her interest in probing the tension between cyberspace and physical space, which led to years-long research into the expression and repression of gender and social inequalities. Initially commissioned by curator John Hanhardt (who was working at the Whitney Museum at the time, but took the project with him to Guggenheim when he joined it in 1998), *Brandon* was set up as a collaborative platform in which artists and curators were invited to respond to these acts of violence, and Brandon's story more specifically. In this sense the project revolved around care in multiple ways: by foregrounding sensitive sociopolitical topics of sexual assault and discrimination and how these were dealt with by police forces and the legal systems. Different organisations and individuals (including artists, curators and general audience members) also cared for the continuation of the work by organising events or adding content to the website. Especially in the years 1998 and 1999, the project started to expand in unexpected directions through the involvement of the various authors and organisations, resulting in installations, online discussion forums, networked performances, and the non-linear website.[12]

Similar to Neddam and Štromajer, Cheang played an important role in the development of the preservation trajectory by explaining the directions that the project and the related presentations took and addressing its sustainability issues. Cheang also mentioned how she regarded *Brandon* as a platform for others to take control of by organising and producing situations that would activate other storylines or collaborators.[13] Although it consisted of many events, the website as the main platform became the best known part of the art project. The website is divided into multiple sections, each with different interfaces – bigdoll, roadtrip, mooplay, panopticon and theatrum anatomicum – that together form the platform. Each interface is programmed as a mainframe: a structural construct in which the contents and collaborators can change. So, while the programming is fixed, the narrative shifts and evolves as a result of new participants as well as technical add-ons and plug-ins. Although users of the site can browse the different sections, the navigation is not straightforward. As Cheang states, it was deliberately created to function as a maze without clear icons or other markers to aid navigation:

> One's ability to investigate, negotiate with the mouse(over) brings different experience of the work. Within a one year stretch, which includes installation, live chat format, actual/virtual performance, no one (including myself) can claim to have viewed the entirety of this work. Pop-up windows on the roadtrip interface, cells of panopticon interface, are all an expansion of the space, spaces to be occupied by various narratives and inhabitants. Surely, non-linear and non-conformative.
>
> *Cheang cited in Ho 2012*

One of the consequences of the intricate and elaborate technical and social network involved in the preservation was that the website malfunctioned. These were both technical, due to software and hardware obsolescence; and social, because of personnel changes at the different organisations. Hence, the website has been offline several times over the years.[14] Matthew Fuller's analysis of the project provides insight into the potential of Cheang's platform and how its organisational allies become part of the aesthetic of collaboration, effectively approximating a network of care:

> Cheang's methods also include creating contexts for the development of artistic languages to emerge. That is to say, she operates at the level of collective individuation in which art and the consideration of its adequacy to the present can be arrived at. Such work implies that there is also an aesthetic of collaboration to be found – an activity core to her work – for instance, in the creation of common platforms or in the curation of the work by other artists, technologists, and musicians with whom she works. Such platforms also establish a condition in which duration begins to operate as a dimension where a work unfolds and finds itself, and in which processing the question of the language of a project becomes part of the palpable working method.
>
> *Fuller 2019*

In other words, despite the unstable situation of *Brandon*, which extended from its artistic conceptualisation to its technical and organisational context, would it be possible to translate an "aesthetic of collaboration" into a network of care to preserve the project? If so, do the various actors, i.e., the organisations, individuals and the technical elements all care to the same degree?

In 2015, Guggenheim initiated a collaboration with students from NYU's Department of Computer Science to preserve *Brandon*. Their goal was to revive it as a living art project, while preserving all functional behaviours and aesthetic properties of the work as defined by the original source code. This involved a combination of code migration, hyperlink replacement, database replacement, and HTML tag and frameset migration. In keeping with conservation ethics and standards, all the changes were documented through version control, treatment reporting and code annotation.[15] In 2017, they relaunched the website. Yet, several challenges undermined a restoration of the entire project: while the website is part of the Guggenheim's permanent collection, the collected ephemera of the offline events are not. Moreover, even though *Brandon* could be reconstructed and studied through documentation and other fragments that are in the archives of different institutions (among others, at De Waag in Amsterdam and Fales Libraries & Special Collections at NYU), not everything is properly processed or accessible. In an attempt to form a network of care we had individual discussions with some of the collaborators from the past: Cheang herself, Dragan Espenschied and Michael Connor (Rhizome), Marvin Taylor (Fales Library & Special Collections), Marleen Stikker (De Waag), and Mark Graham (Wayback Machine).[16] While all these institutions have their own expertise, approaches and work cultures, we wanted to know if they were interested in a potential collaboration, keeping in line with some of Cheang's intentions for the project. For instance, would it be possible to bring individual efforts together so that they contribute to the whole? How does one build on someone else's knowledge? How can sharing and access be improved?

The primary challenges to digital preservation were identified during the discussions. Firstly, financial: since most preservation of digital art is not yet institutionalised, each organisation has its own way of securing funding or allocating budgets. Lacking fixed resources means that most initiatives are project-driven and thus preservation only happens when there is an immediate concern. For instance, as Guggenheim recalled, an earlier effort to preserve *Brandon* was instigated by a request from another museum for a loan, which made them look closer into the functioning of the project (Engel et al. 2018). This way of working is commonly referred to as "conservation-in-action" (Wielocha 2021). Secondly, the reliance on individual efforts: most preservation endeavours are dependent on a specific person, for instance, the artist(s), a curator, or a conservator. Attention for a certain project often lapses when staff are replaced, and consequently specific knowledge and expertise disappear. Moreover, since most institutions don't have a digital art conservator, they rely on external knowledge, which makes it harder to build on past experiences, particularly when decisions are not well documented. Thirdly, the issue of scattered elements: besides the problem of technical obsolescence, with a distributed project such as *Brandon* sometimes parts of the project are lost because it is unclear where or in which institution they are kept. Fourthly, the paradox of digital preservation: having to continuously update the technology to maintain its functioning or aesthetics impedes the restoration or migration efforts. This becomes a technical rat race, in which technical solutions are endlessly stacked on top of each other. In the end it does not only require preserving the project but also preserving the ever-mutating technical environment that is needed to keep the software and hardware functioning.[17]

In conclusion, we noticed how the problem of continuous technical updates doesn't only encumber the project, but also the organisational efforts. Indeed, the Guggenheim has explicitly expressed a need for a dedicated person to lead digital preservation processes, and has underscored the urgency of raising more awareness about digital preservation in the rest of the organisation (Dover 2016). While these challenges are hard for a single institution to overcome, they could be solved by stronger collaborations, i.e., networks of care, in which budgets and expertise are shared, as mentioned by De Waag: "Our starting point is to be situated in the 'art of combining', in the interdisciplinary field. That you are at the edge of your capabilities, and you learn to accept that you are not always the expert."[18] Moreover, it is by acknowledging – and following the art project's aims or characteristics – that preservation happens via various elements and actors that are continuously (re)arranged, following Cheang's wish to keep *Brandon* growing by "commission[ing] artists to expand the interface, like forking out with more episodes, more story development."[19] One way to safeguard such evolvement is by focusing on the relational arrangement of care in which preservation is negotiated between different actors, including humans (the artists, curators, conservators, users and others), as well as on the material and technical elements (including software, hardware and the documentation systems that are used), while accepting that these components may change over time. Finally, recalling Mol et al., (2010) change is not achieved by controlling these elements but should be seen as inherent in the elements and hence, care is temporal and continuous and occurs through experimentation, adaptation and mutation. In other words, such an approach follows the characteristics of a network of care as outlined above.

A Network of Care, or Preservation as an Evolving Process

Referring to shared resources and goals, a network of care includes social relations and negotiations that are necessary to produce and maintain a network and a project. As a model of shared knowledge, it means that not one person has all the information, nor all the power, since the different elements and expertise are distributed. In other words, everyone may own part of a project but the network governs the whole. In line with Fuller's suggestion of an "aesthetics of collaboration," a technical platform can function as a binding element, keeping social relations and potential technical elements together, for example, when (parts of) a project are also archived on the platform. Moreover, the technical construction of the platform can inform the specific information exchange and the ability to follow historical changes. As a consequence, the platform will co-determine the success of a network. This way, it can be argued that technology also cares.

Instead of focusing on specific material elements of a project or on a particular outcome, such an approach regards digital preservation as an ongoing cyclical and evolving process in which various carers come together, share their ideas, but also disperse, reconvene and change, potentially *ad infinitum*. This includes acknowledging that in addition to the actions of humans, materials and technology intrinsically affect the art project as well as the preservation method. Taken together they can offer new perspectives on preservation thinking and doing. Digital preservation as a relational network of humans, materials and technology is executed, reacted upon and consequently evolves or mutates, making it a complex process riddled with kinks, folds, hiccups and slippages, which twist and bend in various directions, creating uncertainty, unpredictable behaviour and surprising results. In this sense, digital preservation can be understood as a speculative practice, where knowledge unfolds between subjects (human and non-human) whose ability to know is mediated by how they reach out, and by the receptivity of the other. Digital preservation then becomes an intriguing combination of adaptability and perseverance, and is formed and developed by the network, in which social, political, economic and technical relations overlap in various ways. In the process the project as well as the network will likely change and can produce new forms of care. While this proposition has the potential – or may seem – to disrupt the status quo, it is not merely about changing or choosing for one or the other option; rather, what the conceptual framework of a network of care proposes is developing a process of relation-making and supporting shared-learning.

The examples I have explored above illustrate new modes of active engagement and creative use, and demonstrate an engaged way of dealing with circulation and sociotechnical relations, in which the distributive effects are intentional even if the outcomes are unforeseeable. Or, paraphrasing Puig de la Bellacasa, a network of care proposes a practice constantly done and undone through encounters that accentuate both the value of trust as well as the awareness of alterity. Moreover, the open and shared process requires a situated ethicality to enable an effective and accountable decision-making process that ensures a more resilient digital preservation practice (Puig de la Bellacasa 2017, p. 115). This means that digital preservation is about striking a delicate balance between care, dependency and equity, in which it is important to continuously question the place of care within or beyond notions of power and ethics as well as the relationships between different dimensions of care.

Acknowledgement

I would like to thank Karin de Wild for collaborating on the case study of Shu Lea Cheang's *Brandon*, and Aga Wielocha and Marina Valle Noronha for the discussions around networks of care.

Notes

1 This is not to imply that the proposed framework can be generalised to become representative of networked culture: instead, the concept of network of care offers an example of best practice, of reference and of comparison to other situations in which humans and non-humans form a relational network.
2 For more information about how I describe the characteristics of net art, see Dekker (2018 pp. 19–33).
3 In her monograph, Mol (2008) describes living with and treating diabetes, and 13 studies from different areas are assembled in the co-edited volume with Mol et al. (2010) positioning the ethics and politics of care.

4 A discussion of "network" is beyond the scope of this chapter, yet importantly, while Deleuze and Guattari emphasise the dynamism and hence the temporal (or the becoming of components) within their thinking, I assume that some entities will also be relatively stable. Hence, the network is not necessarily always based on equality; at certain moments specific components may have more agency than others. Yet, a thorough understanding of all the components and their relations is necessary to understand these dynamics. It is the tension between them that underlies preservation practices.

5 In Dekker (2018 pp. 88–92), I describe the context of networks of care in more detail, in particular, by building on Hui and Halpin (2013).

6 For a more in-depth description of the project and its preservation dilemmas, see Dekker (2018).

7 For more information about the construction of identity through image(s) in mouchette. org, see Warren-Crow (2014).

8 Interestingly, the cease-and-desist letter was addressed not to Neddam (who only outed herself as the author of mouchette.org in 2010), but to mouchette.org, effectively making the website a legal identity.

9 Here there is much to learn from conventional preservation practice where systematics of version control and decision-making models have been developed. See, for instance, Engel and Wharton (2017) and Barok et al. (2019); in the latter a thorough analysis is provided about the influence of a technical system on the way the content can be preserved and understood.

10 Annick Bureaud, www.facebook.com/intima/posts/144916102244400.

11 Personal interview with Shu Lea Cheang, 19 April 2019.

12 For an overview of the different parts of the project, see Engel et al. (2018), and de Wild (2019).

13 Personal interview with Shu Lea Cheang, 19 April 2019.

14 The project was partly funded by Banff in Canada (1995), the Guggenheim in New York (1998), Waag Society in Amsterdam (1997–99), and Harvard University (1999). Over the years several organisations tried to keep the project functioning or archive it, among others, Rhizome and the Internet Archive. For a timeline of the periods of activity and non-activity, see Engel et al. (2018), de Wild (2019).

15 For an elaborate account of the preservation process see Engel et al. (2018).

16 Mark Graham and the Wayback Machine, were never involved in *Brandon's* development; however, over the years they crawled the website and stored screenshots on the Wayback Machine. We believed that this documentation could be relevant to understanding the history – and potential future – of *Brandon*.

17 These challenges are in line with art preservation more generally; however, the speed of deterioration around many aspects of digital art – and thus the need for solutions – is more urgent here. For more information see, among many others, Dekker (2022) and Rinehart and Ippolito (2014).

18 Personal interview with Marleen Stikker, director Waag Society, Amsterdam, 19 May 2019.

19 Personal interview with Shu Lea Cheang, 19 April 2019.

Bibliography

Barok, Dušan, Julie Boschat Thorez, Annet Dekker, David Gauthier and Claudia Röck. 2019. "Archiving Complex Digital Artworks." *Journal of the Institute of Conservation* 42 (2): 94–113. DOI: 10.1080/19455224.2019.1604398

Bhowmik, Samir. 2019. "Thermocultures of Memory." *Culture Machine* 17: 1–20.

Black, Patricia. 2020. "Can 'Mouchette' Be Preserved as an Identity?." *LIMA*, www.li-ma.nl/lima/sites/default/files/LIMA_Can%20%E2%80%9CMouchette%E2%80%9D%20be%20preserved%20as%20an%20identity.pdf.

Cubitt, Sean. 2016. *Finite Media: Environmental Implications of Digital Technologies*. Durham: Duke University Press.

Dekker, Annet, ed. 2010. *Archive 2020: Sustainable Archiving of Born Digital Cultural Content*. Amsterdam: Virtueel Platform.

Dekker, Annet. 2011. "How to Be Pink and Conceptual at the Same Time. Annet Dekker in Conversation with Martine Neddam." In *Because I'm an Artist Too. . . .*, edited by Annet Dekker and Martine Neddam. Amsterdam: SKOR Foundation for Art and Public Domain, pp 22–5.

Dekker, Annet. 2012. "Composting the Net. An Interview with Shu Lea Cheang." *SKOR NetArtWorks*, May 28, 2012. http://aaaan.net/shu-lea-cheang-composting-the-net/.

Dekker, Annet. 2015. "Networks of Care, or How Museums Will No Longer Be the Sole Caretakers of Art." In *DRHA2014 Conference*, edited by Anastasios Maragiannis. London: University of Greenwich, pp 81–5.

Dekker, Annet. 2018. *Collecting and Conserving Net Art. Moving Beyond Conventional Methods*. London: Routledge.

Dekker, Annet. 2022. "Curatorial Perspectives on Collecting Time-Based Media Art." In *Time-Based Media Art Handbook*, edited by Deena Engel and Joanna Phillips. London: Routledge.

Dekker, Annet and Patricia Falcão. 2017. "Interdisciplinary Discussions about the Conservation of Software-Based Art. Community of Practice on Software-Based Art." *PERICLES*, March, 31. 2017.

Deleuze, Gilles and Félix Guattari. 2004[1980]. *A Thousand Plateaus*. Trans. Brian Massumi. London and New York: Continuum.

Depocas, Alain, Jon Ippolito and Caitlin Jones, eds. 2003. *Permanence through Change: The Variable Media Approach*. New York: The Solomon R. Guggenheim Foundation & Montreal: Daniel Langlois Foundation for Art, Science and Technology.

De Silva, Megan and Jane Henderson. 2011. "Sustainability in Conservation Practice." *Journal of the Institute of Conservation* 34 (1): 5–15.

de Wild, Karin. 2019. "Internet Art and Agency. The Social Lives of Online Art." PhD diss., University of Dundee.

Dover, Caitlin. 2016. How the Guggenheim and NYU Are Conserving Computer-Based Art – Part 1." *Guggenheim Blogs*, October, 26. 2016. www.guggenheim.org/blogs/checklist/how-the-guggenheim-and-nyu-are-conserving-computer-based-art-part-1.

Engel, Deena and Glenn Wharton. 2014. "Reading between the Lines: Source Code Documentation as a Conservation Strategy for Software-based Art." *Studies in Conservation* 59 (6): 404–15.

Engel, Deena and Glenn Wharton. 2017. "Managing Contemporary Art Documentation in Museums and Special Collections." *Art Documentation Journal of the Art Libraries Society of North America* (2) (Fall 2017): 293–311.

Engel, Deena, Laura Hinkson, Joanna Phillips and Marion Thain. 2018. "Reconstructing Brandon (1998–1999): A Cross-Disciplinary Digital Humanities Study of Shu Lea Cheang's Early Web Artwork." *Digital Humanities Quarterly* 12 (2). www.digitalhumanities.org/dhq/vol/12/2/000379/000379.html

Fitzpatrick, Kathleen. 2011. *Planned Obsolescence: Publishing, Technology, and the Future of the Academy*. New York: New York University Press.

Fuller, Matthew. 2019. "Inhabiting High Density Realities, on Shu Lea Cheang's Artistic Language." In *3x3x6 Shu Lea Cheang*, edited by Paul B. Preciado. Taipei: Taipei Fine Arts Museum, pp. 15–68.

Gabrys, Jennifer. 2011. *Digital Rubbish*. Ann Arbor, MI: University of Michigan Press.

Harrison, Rodney. 2015. "Beyond 'Natural' and 'Cultural' Heritage: Towards an Ontological Politics of Heritage in the Age of the Anthropocene." *Heritage and Society* 8 (1) (2015): 24–42.

Ho, Yin. 2012. "Shu Lea Cheang on Brandon." *Rhizome*, May 10, 2012. https://rhizome.org/editorial/2012/may/10/shu-lea-cheang-on-brandon/

Hodge, Gail. 2000. "Best Practices for Digital Archiving: An Information Life Cycle Approach." *D-Lib Magazine* 6 (1). www.dlib.org/dlib/january00/01hodge.html.

Hui, Yuk and Harry Halpin. 2013. "Collective Individuation: The Future of the Social Web." In *Unlike Us Reader. Social Media Monopolies and Their Alternatives. INC Reader 8*, edited by Geert Lovink and Miriam Rasch. Amsterdam: Institute of Network Cultures, pp 103–16.

Kagan, Sacha. 2011. *Art and Sustainability. Connecting Patterns for a Culture of Complexity*. Bielefeld: Transcript Verlag.

Laurenson, Pip and Vivian van Saaze. 2014. "Collecting Performance-based Art: New Challenges and Shifting Perspectives." In *Performativity in the Gallery: Staging Interactive Encounters*, edited by Outi Remes, Laura MacCulloch and Marika Leino. Bern: Peter Lang, pp 27–41.

Mol, Annemarie. 2008[2006]. *The Logic of Care. Health and the Problem of Patient Choice*, translated by Peek Language Service and the author. London/New York: Routledge.

Mol, Annemarie and Anita Hardon. 2020. "Caring." In *Pragmatic Inquiry: Critical Concepts for Social Sciences*, edited by John R. Bowen, Nicolas Dodier, Jan Willem Duyvendak and Anita Hardon. London: Routledge, pp 185–204.

Mol, Annemarie, Ingunn Moser and Jeannette Pols. 2010. "Care: Putting Practice into Theory." In *Care in Practice: On Tinkering in Clinics, Homes and Farms*, edited by Annemarie Mol, Ingunn Moser and Jeannette Pols. Berlin: Transcript, pp 7–27.

Nowvisky, Bethany. 2019. "Libraries, Museums, and Archives as Speculative Knowledge Infrastructure." In *Old Traditions and New Technologies: The Pasts, Presents, and Futures of Open Scholarly Communications*, edited by Martin Eve and Jonathan Gray. Cambridge, MA: MIT Press. doi: 10.4324/9780429506765

Pendergrass, Keith L., Walker Sampson, Tim Walsh, and Laura Alagna. 2019. "Toward Environmentally Sustainable Digital Preservation." *The American Archivist* 82 (1): 165–206.

Pope, Kaitlin. 2017. *Understanding Planned Obsolescence: Unsustainability through Production, Consumption and Waste Generation*. London: Kogan Page Ltd.

Prelinger, Rick. 2019. "Archives of Inconvenience." In *Archives*, edited by Andrew Lison, Marcell Mars, Tomislav Medak and Rick Prelinger. Minneapolis, MN: University of Minnesota Press, pp 1–45.

Puig de la Bellacasa, María. 2017. *Matters of Care: Speculative Ethics in More Than Human Worlds*. Minneapolis, MN: University of Minnesota Press.

Rechert, Klaus, Dragan Espenschied, Isgandar Valizada, Thomas Liebetraut, Nick Russler and Dirk von Suchodoletz. 2013. "An Architecture for Community-Based Curation and Presentation of Complex Digital Objects." In *Digital Libraries: Social Media and Community Networks, 15th International Conference on Asia-Pacific Digital Libraries, ICADL 2013, Bangalore, India, December 9–11, 2013. Proceedings*, edited by Shalini R. Urs, Jin-Cheon Na and George Buchanan. Cham: Springer, pp 103–12.

Reinders, Arjan. 2019. "Stokje Eraf, Stokje Eraan." *De Groene Amsterdammer* 51–2, December, 18. 2019. www.groene.nl/artikel/stokje-eraf-stokje-eraan.

Rinehart, Richard and Jon Ippolito. 2014. *Re-Collection: Art, New Media, and Social Memory.* Cambridge, MA: MIT Press.

Sakrowski, Robert and Igor Štromajer. 2017. "EXPUNCTION. Deleting www.intima.org Net Art Works. A Conversation." In *Lost and Living (in) Archives. Collectively Shaping New Memories*, edited by Annet Dekker. Amsterdam: Valiz, pp 159–72.

Summers, Ed. 2020. "Appraisal Talk in Web Archives." *Archivaria* 89 (Spring): 70–102.

Tansey, Eira. 2015. "Archival Adaptation to Climate Change." *Sustainability: Science, Practice and Policy* 11 (2): 45–56.

Van de Vall, Renée. 2018. "Doing Ethics in Conservation Practice: An Example from the SBMK." SBMK Summit on (Inter)national Collaboration. Acting In Contemporary Art Conservation. Amersfoort, November 15, 2018.

Van Saaze, Vivian. 2012. "The Ethics and Politics of Documentation. On Continuity and Change in the Work of Robert Smithson." In *Robert Smithson – Art in Continual Movement: A Contemporary Reading*, edited by Ingrid Commandeur and Trudy van Riemsdijk-Zandee. Amsterdam: Alauda Publishers, pp 63–84.

Warren-Crow, Heather. 2014. *Girlhood and the Plastic Image.* Hanover, NH: Dartmouth College Press.

Wharton, Glenn. 2011. *The Painted King: Art, Activism, and Authenticity in Hawai'i.* Honolulu, HI: University of Hawai'i Press.

Wielocha, Aga. 2021. "Collecting Archives of Objects and Stories: On the Lives and Futures of Contemporary Art at the Museum." PhD diss., University of Amsterdam.

Zavala, Jimmy, Alda A. Migoni, Michelle Caswell, Noah Geraci and Marika Cifor. 2017. "'A Process Where We're All at the Table': Community Archives Challenging Dominant Modes of Archival Practice." *Archives and Manuscripts* 45 (3): 202–15.

11

BEYOND THE SCREENSHOT

Interface Design and Data Protocols in the Net Art Archive

Lozana Rossenova

Introduction

Over the past few decades, the proliferation of improved and optimised scanning technologies, digital record-keeping systems, as well as access to a global market of cheap (often crowdsourced) manual labour, have led to the exponential growth of mass-digitisation and the launch of various online archives and collections (Ridge 2014; Terras 2011; Thylstrup 2019). Cultural heritage institutions, and occasionally large technology companies, have developed policies and procedures to account for the shift in what contemporary cultural stewardship entails (Microsoft 2019; Parry 2010; Parry et al. 2018; Sood 2021). A brick-and-mortar physical storage facility with cataloguing system for access, is no longer enough. Collections of images and texts need to be fully accessible to fulfil the mission statements of increasing number of large national institutions (Kapsalis 2016; Maher and Tallon 2018; Stinson 2017). In the vast majority of cases, access implies access to scanned text pages or high-quality photographs of fine art works and sometimes 3D renderings of sculptural pieces, alongside appropriate, standardised metadata. But cultural heritage is no longer comprised solely of analogue objects, which can be captured via 2D- or 3D-imaging technologies. At the same time, the policies, procedures and attendant digital infrastructures developed by cultural heritage institutions to address the needs of *digitised* collections are not able to address the needs of complex, non-linear and networked *born-digital* cultural expressions, such as multiplayer online video games, user interactions via social media platforms or the case study examined in this chapter: net art.[1] The question addressed in this chapter is whether the screenshot can operate as a representative image of born-digital artefacts, i.e. net art, equivalent to the scan or photograph of a painting? If not, what are the implications for representing and contextualising net art within a digital archival system? While the parallel between screenshots and scanned images may

DOI: 10.4324/9781003095019-16

be perceived as helpful in terms of standardising certain metadata ontologies, this chapter shows the limits to this analogy for net art and, by implication, born–digital cultural works more generally.

The relation between the screenshot and the artwork it attempts to represent is closely tied with the specificity of screen capture as an image-making process and the medium specificity of net art. Unlike other 2D- or 3D-image-making techniques, a screenshot is not representational of any analogue, *digitised* object, but rather captures the result of computational processes rendering what users see and interact with on a screen. A screenshot can be created manually – by user interactions; or programmatically – by the pre-programmed actions of a script intended to perform the screen capture operation at certain intervals or within certain environments. The processes of creation, subsequent execution and final presentation require software – of a specific version, executed within a specific operating system. These get encoded implicitly (at least to the human eye) in the metadata embedded within the resulting screenshot file; and can also be made explicitly visible, if elements of the software (e.g. a browser window frame) or of the operating system (e.g. a scrollbar or a mouse pointer) are captured intentionally, or accidentally, within the section of the screen being captured. Like all other digital images, screenshots can be shared across network protocols, stored in datasets and (re)used for various purposes. However, the immediacy of the capture within the field of a single screen – be it desktop, laptop or, increasingly, the high-quality retina displays of mobile phones – with no need to reference an external object or environment, affords screenshots a particularly important role within digital culture (Gaboury 2019; Švelch 2020). This flattening between the field of capture (the screen), the instrument of capture (software) and the presentation environment (screen software) is what aligns screenshots so closely with net art, making them appear to be the logical, if not even "natural," means to capture and represent such artworks.

Net art works, too, are created, executed and presented via computational environments. However, these works are not single digital artefacts, but rather processual, networked assemblages (Dekker 2018), dependent on alignments between hardware, software environments and network protocols, to be executed and rendered. They change over time and require specific user input in order to be performed (Paul 2009). Furthermore, net art, as described by Michael Connor, Artistic Director of Rhizome, is not just about the creative use of the net but also about examining the conditions of participation in it (Connor 2019). In that sense, it can involve performative, participatory and networked elements outside a browser window. Thus, static screenshots of net art do not behave like the born-digital equivalent of photographs of paintings or 3D scans of sculptures. A single screenshot of a net artwork could be iconographic, i.e. including elements recognisable as part of a specific artwork or artist's style. For example, a screenshot with a repeating image background of close-up flowers, flies and other insects, the portrait photo of a young girl and some default button elements, is a recognisable representation of the net art piece *Mouchette* by Dutch artist Martine Neddam (Figure 11.1). However, just like a single film frame might become an iconic symbol for a film but could not

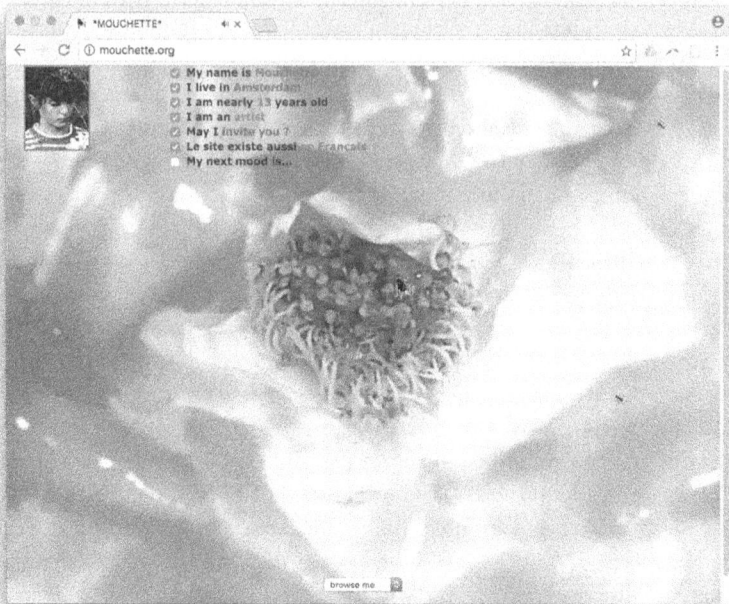

FIGURE 11.1 A screenshot of the landing page of mouchette.org. Image courtesy of Rhizome. [Screenshot: 8 December 2016]

convey the full extent or meaning of the film work, a single screenshot of a net art work cannot convey the full extent of the performative, processual and networked variability of the work. Within this paradigm, a screenshot plays a different role than primary representation – interpreted as a data object, it provides clues to the environment the work was represented in, at the time the screenshot was taken. Elements such as a mouse cursor, scrollbars or even the whole browser interface in cases where that is included in the screenshot, in addition to the metadata encoded within the image file itself, can provide much needed clues as to which environment (and from what time period) can accommodate functional access to a specific artwork. Thus, screenshots play an important role in efforts to archive and preserve net art, but it is not the role typically associated with image-based representations within collection management systems and attendant metadata schemas.

Thinking critically of the potential of the screenshot to capture (or not) and express specific aspects of a net art piece, for example its performative environment, provides a useful framework to outline some of the challenges facing digital archival systems. This chapter considers these challenges in the context of redesigning Rhizome's ArtBase archive of net art (Rossenova 2021). The case study allows one to speak in concrete terms with regards to the architecture of archival software systems, while the chapter maps out the network of relations between operations of the database which structures archival data (the backend) and the interface which enables user interaction (the frontend). Thinking in relational terms opens new strategies for the design of the net art archive and for working with screenshots in archival contexts and by extension contributes to the broader discourse around preserving networked culture.

Current Online Archive and Collection Systems Do Not Account for Processual and Networked Relations

Various initiatives organised by museums and arts organisations have furthered the theory and practice of preserving digital cultural heritage and proposed implementation models for institutional archival workflows – from documentation and acquisition policies to metadata schemas and technical specifications for digital asset management (DAM). In relation to born-digital cultural artefacts, existing research has outlined "significant properties" relating to the conservation of software-based art (Ensom 2018; Laurenson 2014), documentation practices for the purposes of preservation of net art (Dekker 2013) and technical approaches to the preservation of network-based objects (Espenschied and Rechert 2018) and video games (de Vos 2018), to name just a few examples from the current literature. However, multiple questions relating to how these research findings are integrated into the backend database and frontend interface of an archival system remain open.

In the fields of user experience design (UX) and human computer interaction (HCI), research efforts have focused on addressing issues of discovery and accessibility in digital archives. Using digital object "surrogates" (usually image thumbnails and a small selection of visible metadata) in the context of narrative- (Wray et al. 2013) or data-visualisation-based approaches (Whitelaw 2015) have proven influential in moving interface design beyond a purely search-box-based approach – allowing users to interact with digital archives and collections without the explicit need for specialised prior knowledge. Still, most of these approaches are premised on the properties of *digitised* physical objects, such as paintings or books (Kräutli 2016; Vane 2019; Whitelaw 2015; Windhager et al. 2018; Wray et al. 2013), which can be captured and represented via a single, image- or text-based digital file. What is more, strict institutional protocols usually prevent design research projects from deeper engagement with infrastructure – projects tend to remain "client-side" or frontend-focused, i.e. pertaining to what users see in their own browser, which is the client software requesting data from the server software, maintained by the institution. For example, design projects often utilise JavaScript libraries to analyse and visualise data directly in users' browsers in a variety of interesting ways (Whitelaw 2015). Rarely (if ever) are server-side backend database software and data models discussed, or interventions in the underlying data ontologies and organising standards made available to design researchers.

But overemphasising the frontend, or client-side, and the role of the image-based interface misses crucial aspects of what user interaction with net art entails.[2] Complex born-digital artefacts – net art works, video games or social media performances – can prove impossible to summarise, or to extract parameters for meaningful interpretation from, based purely on the visual aspects, i.e. the pixel-layer, of a single, or even several, static screenshot images. The role of image-based representations in contemporary digital media- and online information-sharing as *fixed* or *static*, has already been challenged by media studies scholarship focusing on the performative and operative role of algorithms and networked technologies. As Ingrid Hoelzl and Rémie Marie observe:

> The image as the termination (fixation) of meaning gives way to the image as a network terminal (screen). It is no longer a stable representation of the world, but a programmable view of a database that is updated in real-time. It no longer functions as a (political and iconic) representation, but plays a vital role in synchronic data-to-data relationships. The image is not only part of a programme, but also contains its own 'operation code': it is a programme in itself.
>
> *Hoelzl and Marie 2015, pp. 3–4*

This extension of the conceptual understanding of images into the operative and programmable realm of computation is theoretically better suited to dealing with the question of net art representation and contextualisation, but what does it mean in practice for the archive systems containing images of net art in the form of screenshots? In the process of becoming operational, a screenshot moves beyond traditional, "iconic" representation and accrues metadata, but the metadata needs to be made visible and operational to users in order for them to be able to make meaning of this expanded notion of images. What is more, a screenshot may be a "programme in itself," but it is not the programme the net artist developed; it appears to capture only a still moment, not the performative and processual aspects of the work. In addition, within the context of cultural heritage preservation, there is a persistent "hard" delineation between image files, the (largely analogue) objects they represent, archive records and attendant metadata. Conceptually, digital imaging technology shared via online web interfaces may be *soft* – an enmeshment between software and image-based representation, as per Hoelzl and Marie's defin-ition (2015). But in practice, institutions continue to relegate image files to DAM repositories, wherein fixity checks (procedures that ensure files remain unchanged at bit level) and attendant metadata schemas are designed to preserve images as fixed entities, and *not* as variable, networked data (Arp 2019).

Curators, conservators and researchers who work with net art collections have already noted that existing collection management systems have no provisions to account for the processual, performative and variable properties of born-digital artworks, either in their classification schemas, or in their repository infrastructure or frontend interface design (Barok et al. 2019a, 2019b; Rossenova et al. 2019). Studies of alternative software tools which could begin to address these challenges have focused on documentation and file management (Barok et al. 2019a,b, 2020; Engel and Wharton 2017), but less so on how new ontologies can be organised in flexible backend systems and what their visual representation via a graphical user interface might look like. Finally, existing design metaphors and data visualisation approaches in archival and collection interfaces, such as digital surrogates and virtual white cube galleries (Rossenova 2020b), cannot fully account for the processes involved in the preservation and presentation of the networked and interactive properties of net art. The redesign of Rhizome's ArtBase, discussed next, illustrates the gaps in current digital archive and collection practices when it comes to born-digital culture and leads to a discussion of approaches needed to address these gaps.

Rhizome's Online Archive of Net Art[3]

Rhizome is a grassroots digital arts organisation founded in 1996 as an online mailing list, which then grew into an online platform dedicated to engaging with, promoting and critiquing born-digital art generally, and net art more specifically. In 1999, Rhizome's founder, artist Mark Tribe initiated the ArtBase archive project. The ArtBase invited artists to submit their net art works to the archive and offered various options for data (and metadata) submission and preservation. The archive spans more than 20 years of networked artistic practices and comprises over 2000 artworks offering various ideas of what constitutes a net art archive, its preservation, and how users interact with it. By 2016, the ArtBase archive had already been overhauled several times – both in terms of its backend infrastructure and its frontend design. Mapping out clear links between specific policy decisions, interface designs and previous data model and database implementations, as part of the project to redesign the archive then, highlighted how the problems facing the design and development of born-digital archives were interconnected and interdisciplinary. These include problems concerning the interface design of the archive, the data protocols informing accession, preservation and access policies, as well as the digital infrastructure. This mapping of the problem space expands what may at first seem like a visual design problem: *how can screenshots and other image-based interface elements represent and contextualise net art in the design of the digital archive?;* towards multiple other fields – from HCI to archival science, conservation and curatorial practice, as well as information science and software studies.

Interface Design

In 2016, the landing page of the archive was well integrated into Rhizome's main website. It offered a single entry point into the archive via a grid of small thumbnail images, each standing in for an artwork in the archive (Figure 11.2). Once a user clicked on a thumbnail "surrogate," they accessed a landing page for the artwork record. There was a single screenshot image, presumably from (a section of) the work's interface, a short text description of unclear provenance and only the name of a (single) artist and a (single) date. This artwork representation was not able to answer important contextual questions such as: Who was involved in the artwork production – maybe more collaborators than a single artist? How the

FIGURE 11.2 View of an artwork page in the ArtBase prior to the redesign: includes only artwork title, date, artist name and a short non-structured text description as metadata. [Screenshot: 9 November 2017]

work evolved over time? Did the artist(s) make intentional changes? Did the work change because of updates in the technical protocols of the web and the browsers used to access the web? Or did the ArtBase team change the work in order to preserve it?

To access the artworks, users could click on a button which would take them to a new location (in a new tab) – sometimes linking to a work held in Rhizome's archival repository; sometimes linking to a location on the artists' own server; and on (frequent) occasion – rendering a 404 page, i.e. a missing link. Even if the link was not broken, the artwork that users would gain access to might be in a very different state compared to when it was added to the archive. Parts of the artwork may be broken, missing or dependent on browser plug-ins no longer supported by contemporary browsers. On some occasions, the work might have been updated by the archive team using a particular preservation technique, but that would not be clear by the single access button. All of these scenarios led to perplexing and frustrating user experiences in the archive.

Rhizome's team were aware that the limitations of the latest iteration of the ArtBase interface impacted the value of the archive as a service to the community of archive users (Connor 2018, cited in Rossenova 2020a, p. 99). Reimagining the ArtBase interface in a way that could better meet community needs therefore required more than a redesign involving surface-level branding and styling. It necessitated considering how the net art archive works and for whom – in what ways could the collection management system meaningfully express the various relations between the frontend interface accessed by users and the various tools and infrastructures deployed by in-house staff to store and preserve artwork data *and* metadata. What is more, the development of more sophisticated methods for preservation of net art works driven by R&D processes within Rhizome's preservation team created the need to link artwork archive records not only to a single screenshot image and a single server location where a copy of the work is stored, but to various environments – launching different processes inside the user's browser. Examples include an emulated[4] environment, meeting the technical and network requirements of the original operating system and/or browser version required to perform the work, or a web archive[5] of the work presented via a replay system such as Webrecorder.[6] All of these access options need more clarification (and classification) than simply a button that states "View artwork" next to a static screenshot and highlight the interdependence between processes of preservation, representation and access.

Data Protocols in the ArtBase

Rhizome's staff have adapted the vision and mission statement of the ArtBase over time, as well as its accession policies, in recognition of the challenges net art poses to traditional perceptions of what constitutes an archival record, how it should be described with metadata, preserved and accessed. At the same time, the standard (meta)data protocols and database infrastructures available in the early years of the archive often failed to keep up with the ambitions of the ArtBase's non-standard policies.

When the ArtBase was originally being set up, Mark Tribe consulted various net artists regarding the framework for the archive and offered a choice of how artists wanted their work to be archived (Tribe and Ptak 2010). If artists wanted to hand over digital files, these would be copied on Rhizome's servers and presented in the ArtBase under a rhizome.org sub-domain. Such works were referred to as "cloned objects" (Fino-Radin 2011). Alternatively, if the artists did not want to supply their source files to Rhizome or there was no straightforward technical capacity to do so,[7] then they could provide some descriptive metadata for the work (artist, title, year, short description) and a link to the artwork's URL hosted elsewhere. These works became known as "linked objects" (Fino-Radin 2011). Initially, this hybrid[8] strategy for accession proved flexible. However, within a few years (and sometimes just months) many of the linked objects had disappeared, leading to 404 pages indicating missing resources.

A decade after establishing the archive, the open submission model of the ArtBase was no longer able to keep up with a rapidly expanding field, in which artists working online were looking to distributed social media platforms as new possible venues for artwork-making (Connor 2016). In 2011, the archive stopped accessioning linked objects (Fino-Radin 2011), and in 2015, submissions to the ArtBase were closed entirely. The primary focus of the preservation team then became building tools to facilitate restoring access to works from the archive which have been inoperable for a long time, as well as tools to enable archiving of new artworks – particularly those dependent on third-party platforms. In this context, the preservation of legacy software and environments became a priority, while accessioning of newer works was temporarily paused (Connor 2016). Policy changes related to accessioning highlight the networked characteristics of artworks in the ArtBase. The location of the works – on Rhizome's infrastructure, a server maintained by the artist(s) or a third-party like a social media platform – as well as their documentation with metadata, informs the relations between backend and frontend that contribute to meaning-making in the archive. Such information could not be read from or written into the screenshot-as-representation paradigm alone, it required new data protocols and infrastructures for preservation, representation and access.

The issue with artworks distributed via third-party platforms is particularly pertinent to questions concerning born-digital images and their role in relation to archival practice. In cases of performances reliant on platforms like Instagram, such as Amalia Ulman's work *Excellences and Perfections* (2014) or Guadalupe Rosales' *Veteranas and Rucas* (2015), the work of art spills over the sets of square images neatly confined within an invisible grid on a mobile phone's screen. It extends across the particularities of the app's interface, the default format or editing options for image uploads, the image uploads themselves as they happen over the course of weeks or months, user comments, likes and shares. Similarly, works requesting data from third-party services, such as JODI's *GEO GOO (Info Park)* (2008), which uses elements from Google Maps, or Aaron Swartz and Taryn Simon's *Image Atlas* (2012–), which draws on local image search engine results in countries throughout the world, require the inclusion of a wide variety of external network protocols within the conceptual and technical reach of the net art assemblage. Rhizome's new focus on preservation tools to represent artworks in the ArtBase, therefore, had to address such challenges of archiving around "fuzzy," rather than clear-cut data and user interaction boundaries.

Since 2016, Rhizome have articulated the development of new preservation strategies and tools within the socio-technical framework of a reperformance-as-preservation paradigm (Espenschied and Corcoran 2016). Reperformance refers both to the technical alignments of software and network protocols needed to execute and render the work in a browser, and the interactions users must enact to engage with and experience the work. It involves providing conditions for the execution of born-digital artefacts in an environment which is the same as or as close as possible to the environment the works were originally presented in. In this way, the paradigm points to the way screenshots originally captured for representation and/or documentation purposes and uploaded to the ArtBase alongside "linked" or "cloned" net art works, need to be extended into fully operational and interactive environments.

By developing new digital preservation tools and approaches which can ensure continued access to historic and contemporary networked environments, Rhizome has also claimed license to draw (intentionally) subjective boundaries around the works (Connor 2020; Espenschied and Rechert 2017). What gets included in the artwork assemblage of hardware, software, network protocols and/or third-party services, is determined by the date a work was created, what software would have been in common use at the time and also what clues can be revealed about the software from existing screenshot documentation. It is also worth noting that in many cases, a full operational environment may not be possible because of external network services outside the scope of the boundary that Rhizome can draw. An example is the work *Blackness for sale* (2001) by artists Mendi and Keith Obadike. In 2001, the work was staged as a tongue-in-cheek performative auction on eBay wherein the artists placed Keith's "blackness" up for sale in the site's "Black Americana" category. The artwork critiqued the entrenched Western, colonial values and classifications framing user interactions within the online marketplace, despite the post-racial aspirations of the early web. The work was restaged as a web archive as part of Rhizome's Net Art Anthology exhibition (2016–2018), but the web archive does not link out to anything beyond the single HTML page representing the work's eBay listing. In this case, it can be argued that this reperformance offers little more to users than a full-page screenshot, at least in terms of context. But it is accessible to screen-readers, as it uses live text, and users can open the page in an actual browser, moving a step closer to the original user experience, compared to viewing it as a single static image.

Rhizome's approach to mapping out the artwork boundary privileges the technical and historical expertise of the archivist/conservator and the end-user experience – largely for convenience and efficiency – over the preferences of the artist(s).[9] But this approach also calls for appropriate attribution in the archive whenever artwork reperformances are included as part of the artwork record. This, in turn, requires new ontologies and models in the archival database that can acknowledge the role of external contributors in the historical evolution of the artwork record. What is more, within the reperformance paradigm, users, too, become critical agents in the meaning-making process, since their active involvement is required in order for the works to be rendered and experienced at all. Considering both archivists and users as co-producers of preservation and sensemaking through reperformance helps frame the ArtBase in particular and by extension digital archives in general as networked environments wherein co-creation and collaboration flourish.[10]

To sum up, in order to remain culturally relevant, the net art archive had to ensure the artworks it represented, whose meaning- and value-production resided in the networked relations and user interactions they engendered, remained functionally and conceptually accessible to users. This led to the need to refocus the programme at Rhizome, stop accessioning works and build new tools and infrastructures that could restore gaps and missing links in the network, provide socio-technical context around the works and store all the needed data within an appropriately networked infrastructure and associated data protocols.

Digital Infrastructure

The initial data structure of the ArtBase followed a "basic web model" (Fino-Radin 2011). It began as a MySQL database structured around a custom taxonomy, devised by Rhizome staff members (Rossenova 2020a, pp. 21–23).[11] The open submission policy meant that many of the key terms used to describe and classify artworks were contributed directly by Rhizome's user community. In an interview from 2013, Tribe pointed to the lack of suitable protocols or vocabularies among standard schemas, which could account for the needs of describing net art, including the ability to specify more than one "author" or "artist" for the work, or the possibility to assign different roles for different active participants in the artwork's creation (or preservation) (Tribe et al. 2013). Subsequent efforts to map and migrate existing metadata to established standards, and to utilise a dedicated collection management system to store metadata (the digital catalogue) and link it to the locations where artwork data is stored (the digital repository), revealed the limitations of systems and standards where digital representations remained tied to analogue collection principles (Rossenova 2020a, pp. 21–27).

The list of diverse techniques and additional customisations applied by Rhizome in the production of the Net Art Anthology exhibition alone (Espenschied and Moulds 2019, pp. 433–444) indicated that a fixed vocabulary of standardised procedures and a limited set of relations between the "catalogue" infrastructure for metadata storage, the "digital repository" storing artwork copies and reperformance environments, were not sufficient in the case of the ArtBase. Crucially for the redesign of the ArtBase, the issue could also be reframed around decision-making: how could decisions around reperformance environments and web archive boundaries be surfaced within the interface of the archival framework so that users could understand the context around the reperformance and act accordingly? Preservation work carried out by Rhizome staff involved subjective decision-making, but it was also a collaborative endeavour with the broader artistic and user communities around the archive, because works relied on resources (including screenshots, descriptive metadata, as well as server-side resources) originally provided and often continuously maintained by the artist(s) and could not be reperformed without the participation of users. However, it was not possible to express this network of interdependent relations within previous instantiations of the ArtBase infrastructure.

To address some of these challenges, in 2015, the digital preservation team at Rhizome decided to move away from standard collection management software systems, which tend to act as siloed catalogues describing an external repository (Rossenova 2020a, pp. 27–29). The idea was to instead explore the newly released Wikibase,[12] a free and open-source software for creating, managing and sharing linked open data (LOD). The logical model of the LOD protocol is a network, and not a hierarchical tree as in many other standard classification systems (Dourish 2014). The network model was more apt for the development of a growing ontology around artworks that could change and evolve through ongoing preservation activities, and various forms of additional documentation – texts, screenshots, web archives and more. The flatness of the LOD software environment blurred the sharp edges between the artwork, its multiple instantiations and their documentation, which suited the hybrid accession policies of the ArtBase well. But it also opened new design problems for the ArtBase interface which needed to be able to clearly communicate the shape of the network (its nodes and relations) to users who might have been more familiar with the hierarchical protocols of standard archive and collection databases.

Post-custodial, Networked Archives – New Approaches to Interface Design and Data Protocols

So far in this chapter, I have argued that archives of net art or other forms of born-digital culture can be understood as networks of interdependent, socio-technical relations. Screenshots play a significant role in these networks but they need to be read as data objects capturing a specific moment in time, rather than faithful representations of artworks. Preserving net art within an online archive requires making networked relations visible and accessible for sense-making and interpretation by users. Such visibility and access need to be designed. While there are various approaches that could be taken towards this objective, this chapter focuses on one specific approach, which expands design and HCI methodologies with concepts and theories from the social and archival sciences.

Since articulating the reperformance-as-preservation paradigm, Rhizome have also expressed a desire to move in the direction of a post-custodial archival paradigm, wherein the user community (which includes the ArtBase artists) takes up an active role in processes of classification, preservation and access. Post-custodial practices were developed alongside the rise of electronic record-keeping and the "rediscovery" of the significance of provenance in archival science theory and practice – i.e. considering "records in context," rather than just describing their content, as core to value- and meaning-production in digital environments (Cook 2007, pp. 401–403, 406–407, 414–415). The emphasis is on the importance of "the context, purpose, intent, interrelationships, functionality, and accountability of the record and especially its creator and its creation processes" (p. 418). In other words, there is a shift in focus from "static cataloguing to mapping dynamic relationships" (p. 416), and in the case of the ArtBase – a shift towards a network-shaped database, consisting of nodes connected through specific protocols. But what are the implications of this shift in terms of Rhizome's Wikibase interface and its approach to cataloguing data? Findings from the history of the ArtBase – its changing policies and previous design decisions regarding frontend interface and backend databases – provided evidence that trying to separate problems of interface and interaction design from the context of the underlying data infrastructure is not a productive approach in the design and development of digital archives. Instead, the archive-as-network requires parallel consideration of: (1) the interface design for interacting with data, (2) the protocols and models for structuring data and (3) the database software for storing data, while making the ensuing relations and interdependencies clear to end users.

Infrastructural Inversion

Paying attention to how classification and preservation activities develop, overlap or diverge across different user community and organisational networks in the history of the ArtBase while aiming to make this cultural and technical context available to contemporary users, align the methods of the ArtBase redesign closely with what social scientists Bowker and Star have dubbed "infrastructural inversion" (1999, p. 34). Infrastructural inversion is the act of making visible the underlying data structures and processes of structuring which keep infrastructures "moving along" (p. 313), but normally remain invisible to users. Applying this concept from social science to the field of design practice proposes that taking a reflective, non-opaque approach towards the design of the data protocols, database architecture and interface visuals of the ArtBase means enacting infrastructural inversion. This aims to create possibilities for greater user involvement and collaboration as part of the networked processes in the archive.

In practice, this approach includes providing more ways for users to browse the new ontology and understand new terms and classification taxonomies, linking out to international authority control vocabularies where relevant. In addition, it includes design proposals that treat images (including screenshots) as data and provide high degree of detail around the technical processes and curatorial decisions involved in the preservation and reperformance of a specific artwork. New visual

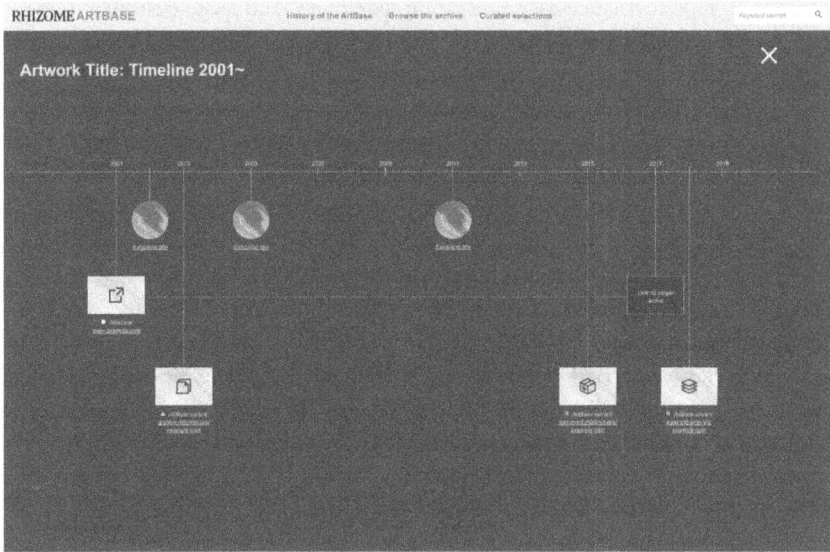

FIGURE 11.3 View of a clickable online prototype for the ArtBase redesign indicating multiple access points to different artwork variants on an artwork record page. [Screenshot: 2 November 2021]

ways for users to browse and discover data nodes and relations across the archive can be facilitated by the non-hierarchical structure of Wikibase and its LOD database. Design visuals such as timelines and multiple, clearly marked access points to different temporal instantiations of a single artwork can be supported by a new overall framework for the data protocols structuring the underlying LOD database (Figure 11.3).

Provenance-Driven Data Modelling

Working through archive policy questions related to access and preservation while defining the objectives for the ArtBase redesign did not immediately lead to fixed design solutions. Instead, it opened up discussions during research sessions with staff and various user communities, which informed the development of new methods for user research and engagement, for organising workshops and utilising design prototypes and, subsequently, devising design specifications. These methods focused on the "mapping [of] dynamic relations" (Cook 2007), such as the role of users in participating (or not) in classification and preservation activities either through direct intervention in the software infrastructure (e.g., through open submission forms) or indirectly – via workshops and community consultation sessions. Various technical and curatorial activities, performed by institutional stakeholders involved in making works available for reperformance via emulation, web-archiving or other preservation strategies, complicated the map of dynamic relations further.

Rhizome's reperformance-as-preservation paradigm and post-custodial policies informed the need to develop a new approach for modelling (meta)data in the archive, which would replace "static cataloguing" (Cook 2007).

The overarching conceptual principle of this approach is not any particular metadata standard, but the principle of archival provenance. The term *provenance* as the summative expression of a data model that can represent the pluralistic context and history around (meta)data in the ArtBase. The term is appropriate if an expanded notion of provenance is taken into consideration as advocated by postmodern archival science scholars Brien Brothman (1991), Terry Cook (2001) and Geffrey Yeo (2013), among others, wherein the archival record is no longer understood to be a static, value-neutral entity, but rather a dynamic process of production and interpretation, carried out by multiple agents. The ArtBase redesign showcases a practical LOD application of the conceptual principles embodied in an expanded definition of archival provenance and context (Rossenova et al. 2019). This new provenance-driven data modelling approach is able to account for changes in a work's technical or user-interaction make-up due to an active intervention by the artist, or due to a contextual event outside the control of the artist, such as a component becoming obsolete. If a preservation action, such as the addition of a web archive or emulation instance of the artwork, has been carried out, this can also be included alongside information about the agent who carried out the preservation action. In short, the focus moves away from the description of siloed objects towards establishing clear patterns for documenting and making visible the relationships between concepts, entities and agents, all part of the archive record's provenance and the networked context.

Model-Database-Interface (MDI) Framework

The redesign of the ArtBase connects the infrastructural inversion approach to interface design and the provenance-driven data model within a new conceptual and methodological framework for the design of born-digital archives. The Model-Database-Interface (MDI) framework proposes concrete strategies to address the challenges not met by traditional collection management systems and metadata standards (Rossenova 2021). While this framework was developed within the particular context of Rhizome, its net art archive and various grassroots user communities; it is a framework that offers several pathways for organisations facing similar requirements and challenges posed by the increasingly networked, data-rich characteristics of born-digital visual culture. These pathways depend on institutional partners moving towards more equal distribution of agency across various stakeholders and involve:

- an openness to sharing institutional processes with a wider network of agents beyond internal staff;
- integration of reflective design methods across the entire cycle of development of an information system (i.e. not reducing design to the act of refining a graphical interface only after infrastructural decisions have been implemented);

- and lastly, fostering meaningful collaborations with users (i.e. involving user communities in processes of classification and preservation, and expanding access interactions with/in unfamiliar environments such as LOD databases).

Conclusion

This chapter has briefly outlined ways in which born-digital cultural expressions challenge established approaches to cultural heritage archiving and preservation. Established metadata standards, collections management software and end-user interfaces cannot account for the needs of temporally variable and performative digital artefacts with networked technical dependencies and user interaction requirements. Screenshots of these artifacts are not equivalent to scans or photographs of physical objects such as paintings or prints. At the same time, screenshots remain the dominant way of documenting and representing born-digital art in online archives and collections. Using the case study of Rhizome's net art archive, the ArtBase, this chapter looked to the ways in which archival design needs to rethink the image-as-artwork-representation paradigm and expand its scope beyond static representation towards operative, data-rich environments. The screenshot as *softimage* (Hoelzl and Marie 2015), containing valuable data on the performative context of net art is a crucial component of the reperformance-as-preservation paradigm advocated by Rhizome. Within this paradigm, artwork preservation is not guaranteed through technical means alone, but through the possibility for users to access and interact with artworks via environments contemporaneous to the works' creation.

In traditional *analogue-to-digitised* archival contexts, acts of storage, preservation and digital representation via image surrogates are separate processes, administered by separate policies and information systems, due to the clear material divide between analogue source objects and digital scans or photographs. In contrast, within *born-digital* archives, as the case of the ArtBase illustrates, preservation and access are far more interdependent – without constant preservation care, artworks easily become entirely inaccessible. In this context, the interface design and underlying data protocols of the net art archive do not present problems that can be clearly defined within a single field or discourse. The problems of the networked archive, are not just archival science problems, design problems or computer science problems, but problems of relations across diverse entities and agents. How can these relations be documented, described and made accessible? The Model-Database-Interface framework developed through the ArtBase redesign case study argues for recognising and making visible the interdependencies of data modelling, backend database administration and frontend interface design. It offers specific design strategies which draw on concepts and methods from STS, social and archival science, besides traditional HCI and UX design. While in practice, pathways for implementation can vary depending on specific domain and community contexts; invariably, networked archives require modes of custodial care and design that span clear-cut disciplinary and institutional vis-à-vis stakeholder boundaries.

Notes

1 The term *net art* is contentious. It is broader than the earlier net.art, which focused on a specific group of mostly European artists during the mid-to-late-1990s, but narrower than the more general born-digital art, which refers to any artwork relying on computers and networks for its production and performativity. In the archive examined here, the ArtBase, the primary experiential context for the artworks is the internet.

2 Throughout the development of HCI as a discipline (Bødker 2015) and the emergence of "user experience" as a concept and a field of study (Forlizzi and Battarbee 2004), the term user and the notion of user interaction have shifted focus and utility from serving to address concerns of ergonomics initially, towards being the subject of behavioural science studies. In this chapter, users are understood to be subjects "co-constructed with technology" (and the designers of technology) following theories from STS (Science and Technology Studies) and the social sciences, rather than the general HCI field. In this sense, users are neither just independent persons unaffected by technology, nor just complete constructs of the designer's imagination (see Oudshoorn and Pinch 2003). Users are active agents in the network of relations that support computational processes and interface interactions, and their agency can be supported – though opportunities for users to make informed choices, or subverted – through over-designed interaction pathways that limit interaction choices (see also Rossenova 2021)

3 Between 2016 and 2021, the ArtBase archive was the main object of study of a collaborative doctoral project between Rhizome and Centre for the Study of the Networked Image aiming at a complete redesign of the archival frontend and backend system. Select findings from this project are extracted here and used as a prompt to discuss networked images in relation to networked art and archival practice.

4 Emulation as a preservation strategy refers to the ability "to emulate obsolete systems on future, unknown systems, so that a digital document's original software can be run in the future despite being obsolete" (Rothenberg 1999). As part of this strategy, Rhizome's team have co-developed the Emulation-as-a-Service framework. EaaS takes advantage of the development of cloud computing services to deliver pre-configured emulation environments which can be reliably deployed online and accessed via a browser. This makes online access to reperformed artworks possible, without requiring users to download and install additional software. See: http://eaas.uni-freiburg.de/.

5 A web archive here refers to an instance of an artwork presented in WARC format, rather than simply as a set of files submitted by the artists and stored on Rhizome's servers. WARC is a standard file format for web archives, which the Library of Congress defines as: "a method for combining multiple digital resources into an aggregate archival file together with related information." The WARC file also contains metadata related to its capture process. See: www.loc.gov/preservation/digital/formats/fdd/fdd000236.shtml.

6 Webrecorder is an open-source tool built by Ilya Kreymer in collaboration with Dragan Espenschied and maintained by a team of developers and designers at Rhizome until 2019, when the project split up. Rhizome's hosted service is now called Conifer, whereas Webrecorder has become an expanded open source project involving multiple tools. See: https://webrecorder.net/ and https://conifer.rhizome.org/.

7 For example, in the case of a complex server-side setup, or if parts of the work were inaccessible to the artist (institutionally, technically or skills-wise), or in cases when the work was technically anchored to its location via absolute URLs being used (Espenschied 2017, cited in Rossenova 2020a, 11).

8 Curator and academic Beryl Graham (2014) has proposed the concept of the "hybrid mode" of collecting as a productive way of thinking through the problem of defining new media or net art archives. She states that as "new media are both tools for collection management and media from which to make art," i.e. "the means of production is also the means of distribution and exhibition," then "in true new media fashion an archive might contain both art and its documentation" (Graham 2014, 48).

9 Given the scarce amount of data available in the ArtBase for some works, and the tendency to use artistic pseudonyms, in many cases, contacting the artists many years after the accession of their works may not even be possible.

10 This relates to Dekker and Tedone's (2019) concept of "networked co-curation", wherein "the complex interrelated network of dependencies and contexts" of the Web create the condition for co-operation between human agents (curators/users) and non-human entities (interfaces, data protocols and algorithms).

11 MySQL is a commonly used open source relational database management system. It is based on the relational model of knowledge organisation which structures data according to entities and their attributed values (see: https://en.wikipedia.org/wiki/MySQL).

12 Accessible at: http://wikiba.se/.

Bibliography

Arp, Charlie. 2019. *Archival Basics: A Practical Manual for Working with Historical Collections.* Lanham, Maryland: Rowman & Littlefield.

Barok, Dušan, Julie Boschat Thorez, Annet Dekker, David Gauthier, and Claudia Roeck. 2019a. "Archiving Complex Digital Artworks." *Journal of the Institute of Conservation* 42 (2) 94–113. DOI:10.1080/19455224.2019.1604398.

Barok, Dušan, Julia Noordegraaf, and Arjen P. de Vries. 2019b. "From Collection Management to Content Management in Art Documentation: The Conservator as an Editor." *Studies in Conservation* 64 (8): 472–489. DOI:10.1080/00393630.2019.1603921.

Barok, Dušan, Julie Boschat Thorez, and Aymeric Mansoux. 2020. *Publishing as a Strategy for Preserving Artistic Research.*

Bødker, Susan. 2015. "Third-Wave HCI, 10 Years Later – Participation and Sharing." *Interactions* 22 (5): 24–31.

Bowker, Geoffrey and Susan L. Star. 1999. *Sorting Things Out: Classification and Its Consequences.* Cambridge, MA; London: MIT Press.

Brothman, Brien. 1991. "Orders of Value: Probing the Theoretical Terms of Archival Practice." *Archivaria* 32: 78–100.

Connor, Michael. 2016. "Interview." In *The New Curator*, edited by Natasha Hoare et al. London: Laurence King Publishing.

Connor, Michael. 2019. "Net Art's Material: The Making of an Anthology." In *The Art Happens Here: Net Art Anthology*, edited by Michael Connor, Aria Dean, and Dragan Espenschied. New York: Rhizome, pp 5–12.

Connor, Michael. 2020. "Curating Online Exhibitions Part 1: Performance, Variability, Objecthood." *Rhizome Blog.* https://rhizome.org/editorial/2020/may/13/curating-online-exhibitions-pt-1/. Accessed May 25, 2021.

Connor, Michael, Aria Dean, and Dragan Espenschied, eds. 2019. *The Art Happens Here: Net Art Anthology.* New York: Rhizome.

Cook, Terry. 2001. "Archival Science and Postmodernism: New Formulations for Old Concepts." *Archival Science* 1 (1): 3–24. DOI:10.1007/BF02435636.

Cook, Terry. 2007. "Electronic Records, Paper Minds: The Revolution in Information Management and Archives in the Post-Custodial and Post-Modernist Era." *Archives & Social Studies: A Journal of Interdisciplinary Research* 1: 399–443.

Dekker, Annet. 2013. "Enjoying the Gap: Comparing Contemporary Documentation Strategies." In *Preserving and Exhibiting Media Art: Challenges and Perspectives*, edited by Julia Noordegraaf, Cosetta G. Saba, Barbara Le Maître, and Vinzenz Hediger. Amsterdam: University of Amsterdam Press, pp 149–169.

Dekker, Annet. 2018. *Collecting and Conserving Net Art: Moving Beyond Conventional Methods.* Oxon: Routledge.

Dekker, Annet and Gaia Tedone. 2019. "Networked Co-Curation: An Exploration of the Socio-Technical Specificities of Online Curation with Annet Dekker." *Arts* 8 (3).

de Vos, Jesse. 2018. *The Game-Shaped Archive: A Playful, Flexible Approach to the Preservation of Computer Games.* Hilversum: Institute for Sound and Vision. https://publications.beelde ngeluid.nl/pub/634. Accessed May 25, 2021.

Dourish, Paul. 2014. "No SQL: The Shifting Materialities of Database Technology." *Computational Culture: A Journal of Software Studies* 4. DOI:10.7551/mitpress/ 10999.003.0006.

Engel, Deena and Glenn Wharton. 2017. "Managing Contemporary Art Documentation in Museums and Special Collections." *Art Documentation: Journal of the Art Libraries Society of North America* 36 (2): 293–311.

Ensom, Tom. 2018. "Technical Narratives: Analysis, Description and Representation in the Conservation of Software-Based Art." PhD diss., King's College London.

Espenschied, Dragan and Heather Corcoran. 2016. "Performing Digital Culture: Dragan Espenschied in Conversation with Heather Corcoran." In *Electronic Superhighway*, edited by Omar Kholeif. London: Whitechapel Gallery.

Espenschied, Dragan and Lyndsey Moulds. 2019. "Preservation Notes." In *The Art Happens Here: Net Art Anthology*, edited by Michael Connor, Aria Dean, and Dragan Espenschied. New York: Rhizome, pp 433–444.

Espenschied, Dragan and Klaus Rechert. 2017. "Final Performance Report. NEH/DFG Bilateral Digital Humanities Program: Tools & Concepts for Safeguarding & Researching Born-Digital Culture."

Espenschied, Dragan and Klaus Rechert. 2018. "Fencing Apparently Infinite Objects: Defining Productive Object Boundaries for Performative Digital Objects." In *Proceedings of iPRES'18*, Cambridge, MA, USA, September 24–27, 2018.

Fino-Radin, Ben. 2011. "Digital Preservation Practices and the Rhizome ArtBase." http:// media.rhizome.org/artbase/documents/Digital-Preservation-Practices-and-the-Rhiz ome-ArtBase.pdf. Accessed February 24, 2017.

Forlizzi, Jodi and Katja Battarbee. 2004. "Understanding Experience in Interactive Systems." In *Proceedings of DIS2004*, August 1–4, 2004, Cambridge, MA, USA, pp 261–268. DOI:10.1145/1013115.1013152.

Gaboury, Jacob. 2019. "Screenshot or it didn't happen." In: *Fotomuseum Winterthur: Still Searching*. www.fotomuseum.ch/en/explore/still-searching/articles/156303_screenshot_ or_it_didnt_happen. Accessed May 25, 2021.

Graham, Beryl. 2014. "Modes of Collection." In *New Collecting: Exhibiting and Audiences after New Media Art*, edited by Beryl Graham. Farnham; Burlington, VT: Ashgate, pp 29–56.

Hoelzl, Ingrid and Remie Marie. 2015. *Softimage: Towards a New Theory of the Digital Image.* Chicago, IL: The University of Chicago Press.

Kapsalis, Effie. 2016. "The Impact of Open Access on Galleries, Libraries, Museums, & Archives." http://siarchives.si.edu/sites/default/files/pdfs/2 016_03_10_OpenCollections_Public. pdf. Accessed May 25, 2021.

Kräutli, Florian. 2016. "Visualising Cultural Data: Exploring Digital Collections through Timeline Visualisations." PhD diss., Royal College of Art.

Laurenson, Pip. 2014. "Old Media, New Media? Significant Difference and the Conservation of Software Based Art." In *New Collecting: Exhibiting and Audiences after New Media Art*, edited by Beryl Graham. Farnham; Burlington, VT: Ashgate, pp 73–96.

Maher, Katherine and Loic Tallon. 2018. "Wikimedia and the Met: A Shared Digital Vision." *Wikimedia Blog*. https://blog.wikimedia.org/2018/04/19/wikimedia-the-met-shared-digital-vision/. Accessed May 25, 2021.

Microsoft. 2019. "A Digital Renaissance Is Helping Global Audiences Connect with Art." https://inculture.microsoft.com/arts/met-microsoft-mit-ai-open-access-hack/. Accessed May 25, 2021.

Oudshoorn, Nelly, and Trevor Pinch. 2003. *How Users Matter*. Cambridge, MA; London: MIT Press.

Parry, Ross, ed. 2010. *Museums in a Digital Age*. Abingdon: Routledge.

Parry, Ross, Doris R. Eikhof, Sally-Anne Barnes, and Erika Kispeter. 2018. "Development, Supply, Deployment, Demand: Balancing the Museum Digital Skills Ecosystem. First Findings of the 'One by One' National Digital Literacy Project." In *MW18: Museums and the Web 2018*. Vancouver, Canada, April 18–21.

Paul, Christiane. 2009. "Context and Archive: Presenting and Preserving Net-Based Art." In *Net Pioneers 1.0: Contextualizing Early Net-Based Art*, edited by Dieter Daniels and Gunther Reisinger. Berlin: Sternberg Press, pp 101–122.

Ridge, Mia, ed. 2014. *Crowdsourcing Our Cultural Heritage. Digital Research in the Arts and Humanities*. Farnham: Ashgate.

Rossenova, Lozana. 2020a. "ArtBase Archive Context and History: Discovery Phase and User Research 2017–2019." https://lozanaross.github.io/phd-portfolio/docs/1_Report_ARTBASE-HISTORY_2020.pdf%0A. Accessed May 25, 2021.

Rossenova, Lozana. 2020b. "Design Landscape Online Collection Interfaces: Discovery Phase and User Research 2018." https://sites.rhizome.org/artbase-re-design/docs/3_Report_DESIGN_LANDSCAPE_2020.pdf. Accessed May 25, 2021.

Rossenova, Lozana. 2021. "Model–Database–Interface: A Study of the Redesign of the ArtBase, and the Role of User Agency in Born-Digital Archives." PhD diss., London South Bank University.

Rossenova, Lozana, Karin de Wild, and Dragan Espenschied. 2019. "Provenance for Internet Art: Using the W3C PROV Data Model." In *Proceedings of 16th International Conference on Digital Preservation iPRES 2019, September 16–20, 2019 Amsterdam, The Netherlands*.

Rothenberg, Jeff. 1999. *Avoiding Technological Quicksand: Finding a Viable Technical Foundation for Digital Preservation*. Washington, DC: Council on Library & Information Resources.

Sood, Amit. 2021. "Google Arts & Culture Turns 10." *Google Arts & Culture Blog*. https://blog.google/outreach-initiatives/arts-culture/google-arts-culture-turns-10/. Accessed May 25, 2021.

Stinson, Alex. 2017. "How Do Memory Institutions Use Wikipedia and Wikidata in Their Collection Catalogues?" *Wikimedia Blog*. https://blog.wikimedia.org/2017/12/18/wikipedia-wikidata-collection-catalogues/, Accessed May 25, 2021.

Švelch, Jan. 2021. "Redefining Screenshots: Toward Critical Literacy of Screen Capture Practices." *Convergence* 27 (2): 554–69. https://doi.org/10.1177/1354856520950184.

Terras, Melissa. 2011. "The Rise of Digitization." In *Digitisation Perspectives*, edited by Ruth Rikowski, 3–20. Rotterdam: Sense Publishers.

Thylstrup, Nanna B. 2019. *The Politics of Mass Digitisation*. Cambridge: MIT Press.

Tribe, Mark and Lauren Ptak. 2010. "Interview with Mark Tribe." www.as-ap.org/oralhistories/interviews/interview-mark-tribe-founder-rhizome. Accessed May 23, 2017.

Tribe, Mark, Crystal Sanchez, Claire Eckert, Mika Yoshitake, and Lauren Teal. 2013. "Interview with Mark Tribe." *Smithsonian Institution Time-Based and Digital Art Working Group: Interview Project.* www.si.edu/content/tbma/documents/transcripts/MarkTribe_130524.pdf. Accessed May 25, 2021.

Vane, Olivia. 2019. "Timeline Design for Visualising Cultural Heritage Data." PhD diss., Royal College of Art.

Whitelaw, Mitchell. 2015. "Generous Interfaces for Digital Cultural Collections." *Digital Humanities Quarterly* 9 (1). www.digitalhumanities.org/dhq/vol/9/1/000205/000205.html. Accessed May 25, 2021.

Windhager, Florian, Paolo Federico, Günther Schreder, Katrin Glinka, Marian Dörk, Silvia Miksch, and Eva Mayr. 2018. "Visualization of Cultural Heritage Collection Data: State of the Art and Future Challenges." *IEEE Transactions on Visualization and Computer Graphics* 25 (6): 2311–2330. DOI:10.1109/TVCG.2018.2830759.

Wray, Tim, Peter Eklund, and Karlheinz Kautz. 2013. "Pathways through Information Landscapes: Alternative Design Criteria for Digital Art Collections." In *Proceedings of 2013 International Conference on Information Systems*, 1–21. http://aisel.aisnet.org/cgi/view content.cgi?article=1191&context=icis2013.

Yeo, Geoffrey. 2013. "Trust and Context in Cyberspace." *Archives and Records* 34 (2): 214–234. DOI:10.1080/23257962.2013.825207.

INDEX

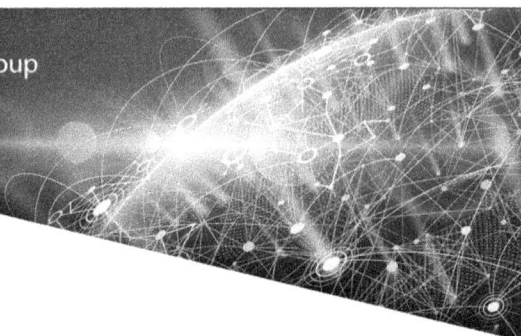

Taylor & Francis Group
an **informa** business

Taylor & Francis eBooks

www.taylorfrancis.com

A single destination for eBooks from Taylor & Francis
with increased functionality and an improved user
experience to meet the needs of our customers.

90,000+ eBooks of award-winning academic content in
Humanities, Social Science, Science, Technology, Engineering,
and Medical written by a global network of editors and authors.

TAYLOR & FRANCIS EBOOKS OFFERS:

A streamlined
experience for
our library
customers

A single point
of discovery
for all of our
eBook content

Improved
search and
discovery of
content at both
book and
chapter level

REQUEST A FREE TRIAL

support@taylorfrancis.com

Routledge
Taylor & Francis Group

CRC Press
Taylor & Francis Group

For Product Safety Concerns and Information please contact our EU
representative GPSR@taylorandfrancis.com
Taylor & Francis Verlag GmbH, Kaufingerstraße 24, 80331 München, Germany

www.ingramcontent.com/pod-product-compliance
Lightning Source LLC
Chambersburg PA
CBHW060252220326
41598CB00027B/4075